**St. Louis Community College**

Forest Park
Florissant Valley
Meramec

Instructional Resources
St. Louis, Missouri

HOW TO READ CIRCUIT DIAGRAMS AND ELECTRONIC GRAPHS

Martin Clifford

Prentice Hall
Englewood Cliffs, NJ 07632

Library of Congress Cataloging-in-Publication Data

Clifford, Martin
    How to read circuit diagrams and electronic graphs / Martin
Clifford.
      p. cm.
    Includes index.
    ISBN 0-13-430802-6
    1. Electronics—Charts, diagrams, etc. I. Title.
TK7866.C55 1988   87-24079
621.381'022'3—dc19   CIP

Editorial/production supervision: Karen Winget
Cover design: George Cornell

 © 1988 by Prentice-Hall, Inc.
A Division of Simon & Schuster
Englewood Cliffs, New Jersey 07632

Printed in the United States of America

10 9 8 7 6 5 4 3 2 1

ISBN 0-13-430802-6

PRENTICE-HALL INTERNATIONAL (UK) LIMITED, *London*
PRENTICE-HALL OF AUSTRALIA PTY. LIMITED, *Sydney*
PRENTICE-HALL CANADA INC., *Toronto*
PRENTICE-HALL HISPANOAMERICANA, S.A., *Mexico*
PRENTICE-HALL OF INDIA PRIVATE LIMITED, *New Delhi*
PRENTICE-HALL OF JAPAN, INC., *Tokyo*
SIMON & SCHUSTER ASIA PTE. LTD., *Singapore*
EDITORA PRENTICE-HALL DO BRASIL, LTDA., *Rio de Janeiro*

To Linda and John Fitzpatrick
with affection

Contents

Chapter 2 PICTORIALS
AND SYMBOLS   **29**

Chapter 3 FUNDAMENTAL
CIRCUITS   **90**

Chapter 4 SUBSCHEMATIC DIAGRAMS 120

Chapter 5 PCs, ICs, AND OP AMPS 155

Chapter 6 PARTIAL AND COMPLETE BLOCK AND CIRCUIT DIAGRAMS 196

Chapter 7   LOGIC CIRCUITS   230

Chapter 8   WAVEFORMS
AND MISCELLANEOUS
CIRCUITS   256

Chapter 9   HOW TO READ
ELECTRONIC GRAPHS   285

Index   **321**

Preface

The study of electronics can be arranged into three categories: basic electronic theory; mathematics for electronics; and a study of circuit diagrams and graphs. It would require an encyclopedic book to cover all these subjects adequately. In any of these topics there is always some crossover. A book on basic electronics theory always contains some circuit diagrams; one on diagrams inevitably includes some theory, and electronic mathematics may have both theory and diagrams.

The purpose of this work is to analyze circuit diagrams in depth with the thought that this will lead to a better understanding of electronics and greater familiarity with the symbols used in making diagrams.

In the years that have elapsed since the early days of radio and television, circuit diagrams have become more complex due directly to the increasing sophistication of in-home electronics entertainment equipment. At one time a radio receiver might have been equipped with three to five tubes. A modern set could contain hundreds of transistors if those contained in integrated circuits are included in the count. At one time it was also simple to trace the path of a signal through a receiver from antenna input to speaker output, but it is no longer as easy to do so.

This book makes no claim that it includes only the latest circuits. Those covered range from old to new, and there are even some that are antiques. Popular circuits seem to appear and disappear and so do electronic components. Early radio sets were solid-state and used crystal detectors. With the arrival of the triode vacuum tube, the crystal detector was relegated to the electronic attic. But when solid state returned in the form of the transistor, the crystal detector, in a somewhat changed form, was put back in its original highly popular position. The individual transistor now shares electronic gear with integrated circuits (ICs).

This book contains numerous electronic symbols, numerous for two reasons. The first is that there has been a substantial increase in available electronic parts, plus variations of these parts. The other is that manufacturers of electronic equipment do not seem bound to follow a set of symbol standards. Fortunately, there is sufficient resemblance among all the symbols so they are recognizable despite their changes.

With all this, being able to draw or at least to read circuit diagrams is important for a good grasp of electronics: it is an aid in learning theory and it is even helpful in electronics mathematics.

The final chapter of this book includes an analysis of graphs used in electronics. These are often used in a study of the behavior of individual parts or circuits and are an excellent way of depicting, graphically, the action of these parts or circuits under a variety of working conditions.

ACKNOWLEDGMENTS

My special thanks to the following companies who provided suitable amounts of data and encouragement, both of which were welcome.

Alpine Electronics of America, Inc.
Channel Master, Div. Avnet, Inc.,
Cherry Semiconductor Corporation
dbx, Inc.
General Electronic
Kyocera International, Inc.
Motorola, Inc.
Programming & Systems, Inc.
Radio-Electronics

Martin Clifford

1

Block Diagrams

Various kinds of diagrams are used to show the relationship of electronic parts or circuits to each other, or to supply a partial or complete graphic description of a component such as a receiver, transmitter, video cassette recorder, compact disc player, or comparable units. Diagrams also show parts values or voltage waveshapes and in some instances the direction of flow of signals or currents. A diagram is an overview and can be regarded as an electronic blueprint.

Electronic diagrams can exist in a number of different forms: block diagrams; circuit, pictorial, and printed circuit board (PC) diagrams; wiring and chassis layout diagrams; and operational amplifier (op amp), construction, chassis layout, and parts placement diagrams. Some are esoteric and rarely used; others are quite common.

Diagrams can be dedicated or integrated. A dedicated diagram is one that is not mixed with any other type. A block diagram using blocks only is dedicated. An integrated diagram is one that utilizes two or more different diagram types.

A block diagram consists of a number of rectangles or squares (or other geometrical figures) arranged horizontally or vertically, or some combination of the two, and connected by short, straight lines. In some instances the lines are terminated by arrows to emphasize the direction of movement of a signal or to indicate the input and output ports of a circuit.

ADVANTAGES OF BLOCK DIAGRAMS

The block diagram is the easiest to draw. It can be done freehand or by using a template and is readily subject to change. It can be made to supply as much or as little data as required and is an excellent method of supplying an overall view of a complete component or some section of it. It also furnishes a good concept of how the various circuits relate to each other.

BASIC BLOCK DIAGRAM

Block diagrams range from a single block to some that are highly involved and quite complex. Figure 1.1 shows the simplest possible block diagram consisting of just an AM receiver. The input from the antenna to the receiver is a single wire. The receiver is monophonic, and a single speaker is often used for this type of receiver.

This diagram is an integrated type of block diagram since it uses electronic symbols for the antenna and speaker with a single block to represent the receiver. Although the antenna and speaker are identified by callouts placed alongside the symbols, these are not always included. However, the block must be identified by data placed inside or alongside it. The antenna is the input; the speaker the output.

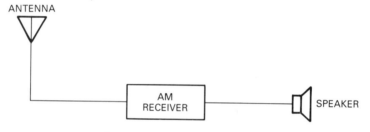

Figure 1.1. Basic block diagram.

LINE LIMITATIONS

The lines that connect the components in a block diagram are often no indication of the number of wires being used. The single connecting line between the FM antenna and the receiver in Figure 1.2 actually consists of a pair of wires. The single line between the AM/FM/MPX receiver and the left and right speaker systems uses eight wires and comprises two pairs of left/right wires for the left speaker system and similar pairs for the other speaker system.

Figure 1.2 is a combined vertical/horizontal block diagram. The main signal flow is from the antenna to the receiver and then to the speakers. The vertical blocks are used for auxiliary or add-on devices, such as (in this case) a compact disc (CD) player and a pair of headphones. The dashed lines around the speakers have no electronic significance and are used to indicate speaker enclosures. The speakers are two-way types, but other than that the drawing supplies no further information.

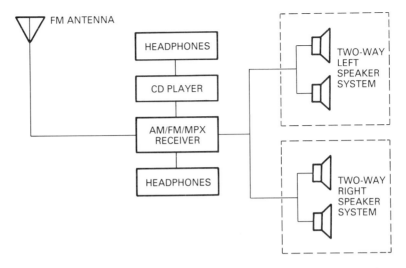

Figure 1.2. Single lines do not indicate the number of connecting wires used.

Block Positioning

In a horizontal diagram, the movement of the signal is usually considered to be from left to right. There may be some indication as to the location of the input and output terminals, but if not, the leftmost block is often the input; the rightmost the output.

Signal flow in a horizontal block diagram isn't always in one direction. In some electronic components either all or some portion of the signal may be fed back to the same or some earlier stage (Figure 1.3). When this is done, arrows are often used to emphasize the action.

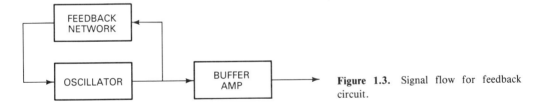

Figure 1.3. Signal flow for feedback circuit.

CONNECTING LINES

The lines joining the blocks are conductors. However, in a block diagram information is seldom supplied as to the number of connecting wires, their size, input and output impedances, or the type of plugs and jacks that are used.

While the lines are easy enough to follow they can sometimes be confusing as in the case of a multiconductor cable. Lines are always drawn so they have the same weight; that is, no line is thicker or thinner than any other. This does not indicate that all the connecting wires have the same gauge.

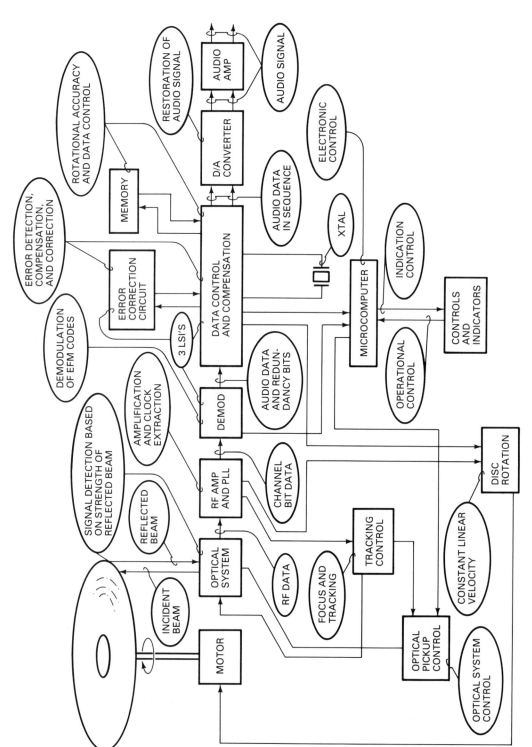

Figure 1.4. Balloons used to indicate circuitry functions.

CALLOUTS

There isn't enough space inside a block to supply detailed data, but generally information is furnished that indicates what the block represents. It may be used to identify a circuit, it can be a manufacturer's part number, or it can indicate some voltage value.

If the area in the block is inadequate, the data can be condensed by using abbreviations or an acronym. If the acronyms aren't in common usage, they can be explained by a footnote.

Balloon Callout

When more information must be supplied than the block can accommodate, a balloon callout (Figure 1.4) can be used. It isn't the purpose of a block diagram to give explanatory details. Block diagrams are an excellent way of learning electronic circuitry, and balloons can be used to emphasize the operational feature of a circuit. Balloons can be used external to the blocks in a diagram or they can be made an integral part of it (Figure 1.5). Each block is used to indicate a circuit; each balloon or circle shows some circuit behavior.

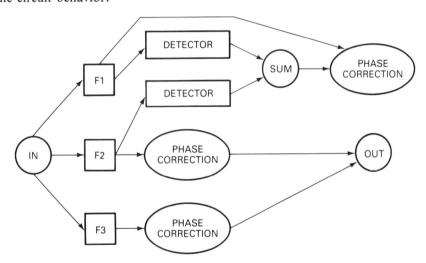

Figure 1.5. Balloons as integral part of block diagram.

DRAWING BLOCK DIAGRAMS

The easiest, but not the most desirable or the neatest, way of drawing a block diagram is freehand. A better and faster method is to use a template. The data inside each block should not be script (often illegible) but in block letters, preferably all capitals. Punctuation isn't necessary, even for abbreviations. For electronics, some templates supply electronic symbols only, but do not have provision for blocks. In that case

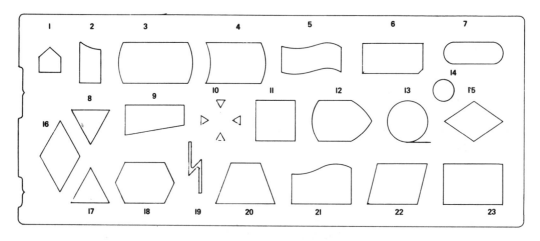

1.	Offpage connector	* 7.	Terminal	13.	Magnetic tape		19.	Communications Link
2.	Transmittal tape	8.	Merge	*14.	Connector	*	20.	Input/Output
3.	Keying	9.	On line keyboard	*15.	Decision		21.	Document
4.	On line storage	10.	Arrowheads	16.	Sort		22.	General Function
5.	Punched Tape	11.	Auxiliary Operation	17.	Extract	*	23.	Processing
6.	Punched Card	12.	Display	*18.	Predefine Process			

Figure 1.6. Each flowcharting symbol represents a specific function. (Courtesy Programming & Systems, Inc.)

use a template (Figure 1.6) made for computer flowcharts. These always supply an outline for a block and for other geometric figures used in block diagrams.

PARTIAL BLOCK DIAGRAM

A block diagram can be complete or partial. A block diagram of a receiver could show all of the important circuits from the signal input point to the speakers and such a diagram would be complete. It isn't necessary to draw an entire diagram if there is interest in just a small section (Figure 1.7). This is a combined vertical/horizon-

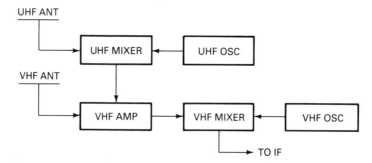

Figure 1.7. Block diagram can be used to indicate portion of the circuitry of a component.

tal diagram. The UHF (ultra high frequency) mixer receives signal inputs from two sources: the UHF antenna and a UHF oscillator. The VHF (very high frequency) amplifier also receives signals from two sources: the VHF antenna and the UHF mixer. In a diagram of this kind, arrows are essential to show signal movement. This block diagram has a feature common to many such diagrams since it clearly designates an input, the UHF and VHF antennas, and an output marked "to IF" (intermediate frequency).

BLOCK SYMBOL VARIATIONS

The squares or rectangles used in a block diagram have no inherent meaning. Thus, a block can be used to represent any one of a large number of circuits. These are identifiable only by the data inscribed in the block.

There are symbols other than rectangles or squares that can be used in a block diagram, but these have a specific meaning and for that reason require no separate identification. Figure 1.8 shows such a diagram using triangles, with each triangle representing an amplifier. No data are put inside the triangles, but information can be supplied outside the symbol. In this drawing, the word "amplifier" is placed above the triangles, but this is unnecessary since the symbols alone identify the circuit as an amplifier. The numbers 20 dB and 40 dB indicate the gain of the amplifiers.

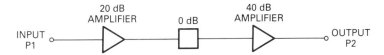

Figure 1.8.   Triangle is the symbol for an amplifier; small circles are used to represent input and output.

INPUT AND OUTPUT POINTS

The input and output points of a block diagram can be indicated by pair of small circles, shown in Figure 1.8. In this case the words "input" and "output" are also used; this is sometimes done either for emphasis or to avoid any possible ambiguity. In some instances the circles are omitted, but in that case written identification is necessary.

The alphanumeric designations, P1 and P2, associated with the input and output points, are simply for written or spoken reference. There is no standardization, and any letters or numbers can be used.

BLOCK GROUND

The word "ground" used in connection with diagrams has two meanings. It can represent an actual connection to earth, and this can be via the ground side of an AC outlet, or a pipe that does go into the ground, such as a cold-water pipe. A ground can also represent a common connection (Figure 1.9A), and in this respect it helps simplify the drawing of diagrams.

At one time electronic components were constructed on a metal chassis, and instead of using printed circuit boards had point-to-point wiring. The chassis was a convenient way of making common connections and was frequently referred to as *ground*. In some instances the chassis had a terminal mounted on the rear apron for connection to an external ground, but often enough this terminal wasn't used.

Since printed circuit boards are made of a nonconductive material, the convenience of the metal chassis was lost. Instead, a conductor often extending the length of the board was used in its place, and this became the common connector. Sometimes, to indicate a common connection without attachment to an external ground, the symbol in Figure 1.9B is used.

Ground is sometimes identified by the abbreviation "gnd" but more often no designation is used. Sometimes the ground connection is one of the terminals of the input as well as the output. In this drawing the input consists of terminal P1 and ground; the output is terminal P2 and ground.

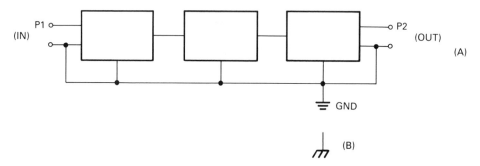

Figure 1.9.   Diagram with input and output clearly designated (A); alternate ground symbol (B).

POWER SUPPLY BLOCK DIAGRAM

The blocks in a diagram are connected by lines with the exception of the power supply, which may or may not use such lines. The power supply is common to all the circuits, and so it is generally understood that a voltage distribution line is connected to them.

The block diagram in Figure 1.10 shows a number of interconnected components, with each using a ground connection. The power supply, either a battery or an electronic type, has a DC output. Using the ground symbol, the negative or ground terminal of that supply is connected to each of the blocks but without the use of lines. The plus terminal (+) of the power supply is also joined to each of the blocks, again without the use of lines. In this case it is assumed that each of the blocks obtains its plus DC voltage from the power supply.

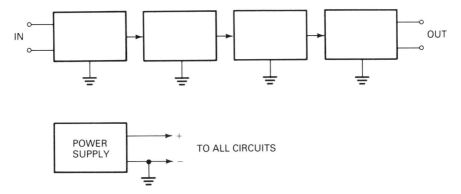

Figure 1.10. Method of showing power supply in a block diagram.

Shunt Voltage Delivery

In some block diagrams, the voltage from the power supply is shown connected as in Figure 1.11A. The advantage of this diagram is that it shows the relationship of all the circuits to the power supply, emphasizing they are all in shunt (parallel) with

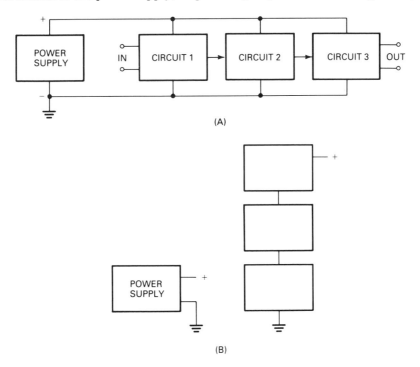

Figure 1.11. Circuits wired in shunt across power supply (A); components working as a series voltage divider (B).

that supply. The advantage of drawing the block diagram this way is that it simplifies the drawing by eliminating a large number of connecting lines.

Series Voltage Delivery

Most often components are shunted across the power supply, but sometimes the components receiving the voltage are wired in series, as in Figure 1.11B, with the components functioning as a voltage divider.

If we consider the three components as the equivalent of a single unit, then they are shunted across the power supply. In either the shunt or series arrangement, the components need not get the same amount of voltage. The voltage may be reduced in the circuit to meet the required demands of that circuit.

DASHED LINES

Dashed lines have a variety of uses in block diagrams. Such lines can be open or closed; if open the line can be of indeterminate length; if closed they can form a box. The lines can represent a box that is simply an enclosure such as that used for speakers or integrated circuits. The enclosure can also be made of metal, and in that case its function is to protect its contents against unwanted magnetic fields. Ordinarily, the

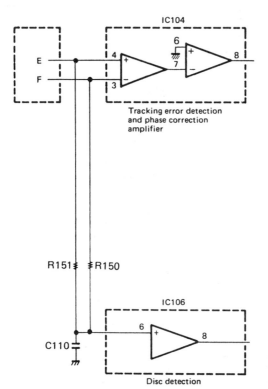

Figure 1.12. Dashed lines used to indicate an integrated circuit (IC). (Courtesy Kyocera International, Inc.)

metal box is grounded as indicated by a ground symbol connected to the box. In some instances, however, that symbol is omitted.

The dashed lines forming boxes in Figure 1.12 are used to indicate an insulating enclosure for integrated circuits (ICs).

A vertical dashed line can be used to separate a pair of blocks that constitute an entire component with the first block controlling the action of the second (Figure 1.13). In this example the control circuit consists of a reed relay requiring an operating current ranging from 1 to 30 milliamperes. The controlled circuit has a current capability of 1.5 to 30 amperes.

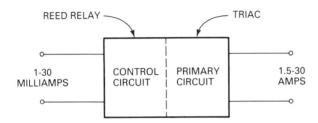

Figure 1.13. Dashed line used to separate a pair of circuits.

Although the control and primary circuits in this drawing are shown as adjacent to each other, it does not automatically follow that they are physically close in the unit; the primary circuit could even be external to the component. This is also true of the circuit diagrams for these two blocks. The advantage of the block diagram in this case, as it is in so many others, is that the electronic relationship of the circuits represented by the blocks is immediately evident. Thus, while a block diagram might be considered by some as an elementary form of diagramming, it is as important and necessary as any other.

WAVEFORMS AND THE BLOCK DIAGRAM

In its movement from an input to an output, a signal will undergo a number of changes. Its amplitude will increase and its shape may be different from its original form. It may be used to trigger another circuit into action for a predetermined amount of time, or it may simply be amplified on a continuous basis. It may undergo a phase change with respect to the input signal (Figure 1.14).

Voltage waveforms are sometimes placed at the input or output side of a block (or both) as an indication of the effect of the circuitry on the signal. The voltage value of the waveform is sometimes included, but the waveforms themselves are rarely drawn to scale. In some instances the waveforms are directly photographed from

Figure 1.14. Voltage waveforms may accompany a block diagram.

the screen of an oscilloscope, reduced in size, and then used in conjunction with the blocks of the diagram.

DIGITAL DISPLAY BLOCK

Many receivers use a tuning scale numbered with the operating frequencies of AM and FM stations. Precise tuning is often difficult even with the help of signal strength and tuning meters. The difficulty is compounded by the fact that the tuning scale is often not at eye level. These problems are overcome by using a digital frequency display (Figure 1.15). This is indicated in a block diagram by a block containing numbers selected at random. If the selected station is on the AM band, four digits are used; if on the FM band, two whole numbers and a single decimal are used.

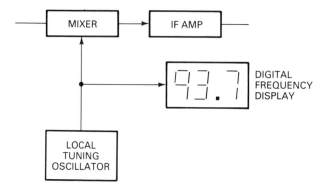

Figure 1.15.   Block diagram showing digital frequency display.

The digital frequency display block has two functions: (1) it calls attention to the fact that tuning is done digitally, and (2) it also indicates the position of the digital circuitry with respect to the other circuits in the component. However, this block supplies no further information as to the type of display that is used.

While a digital frequency display is highly advantageous, not all receivers are so equipped, and many still use a frequency scale across the front. Unlike the digital display, this scale is not electronic and is not part of a block diagram.

OPERATION INDICATORS

Meters, fluorescent bar graphs, and light-emitting diodes (LEDs) are used as indicators of the status of electronic components. In a block diagram indicators can be represented by small circles with lines connecting those circles to the block from which they get their input signal (Figure 1.16).

In this drawing the input signal uses an alternative antenna symbol. The two lines connecting the antenna to the first block indicate twin lead used as the transmission line between the antenna and the receiver input. For AM the transmission line is shown by a single line for the wire between the antenna and the input.

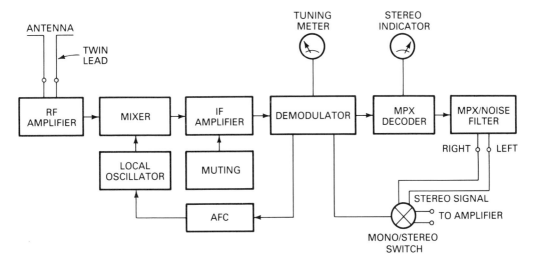

Figure 1.16. Circles are used as operation indicators. The two lines from the antenna represent twin lead.

SIMPLIFYING THE COMPLEX BLOCK DIAGRAM

In some instances, determining the input and the output of a block diagram is quite difficult, especially if these are not positioned near the beginning and end of the diagram. If the diagram is that of a receiver, the input is an antenna and the output is a speaker system, and so even if such diagrams are complex, finding the input and output is easy, and the movement of the signal is fairly straightforward.

But even with these, the block diagram can be quite elaborate because of the presence of a large number of blocks (Figure 1.17A). As a clue, consider that the signal path is usually horizontal, moving from left to right. Add-on equipment often consists of vertically arranged blocks connected by flowlines to the main signal path.

To analyze such a diagram, divide it into three sections as in B, C, and D. This is a fairly obvious method of simplifying the diagram, but there are some that do not yield as easily to this approach.

The diagram in Figure 1.17 handles a single FM broadcast signal. A TV receiver, though, picks up a composite signal that is actually three signals in one: the video signal, the audio signal, and a variety of pulses, including horizontal and vertical sync pulses. For part of the receiver these signals travel together, but they are ultimately separated and follow individual paths.

CONNECTOR BLOCK DIAGRAMS

At one time the only signal source for a TV set was a broadcast TV station. Today, however, that set is the central component of an in-home entertainment system and, in addition to broadcast TV, can accept signals from a video cassette recorder (VCR),

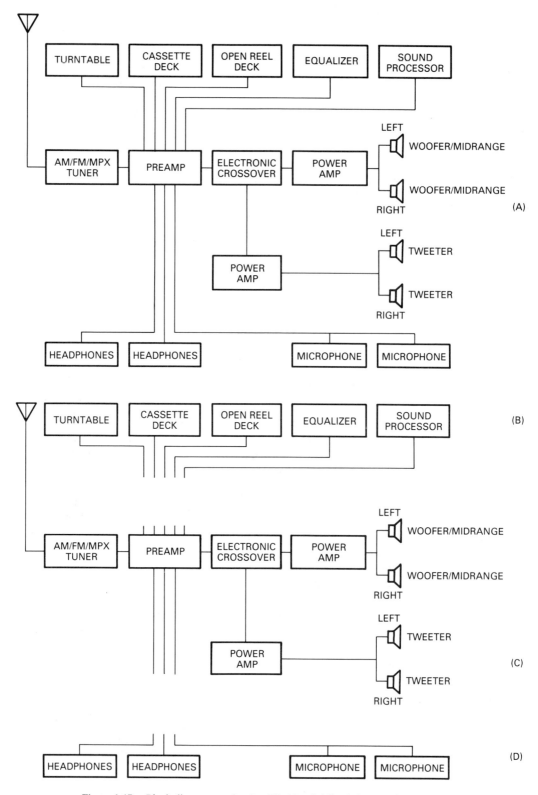

Figure 1.17. Block diagram can be simplified by dividing it into sections as in B, C, and D.

a video game, and subscription or pay TV. The TV receiver has just one pair of input terminals and so there are only two alternatives for using these signal sources one at a time. The first is to connect each of them, as required. This is a nuisance, and if repeated too often, can damage the connectors. A better way is to use a video switcher (Figure 1.18).

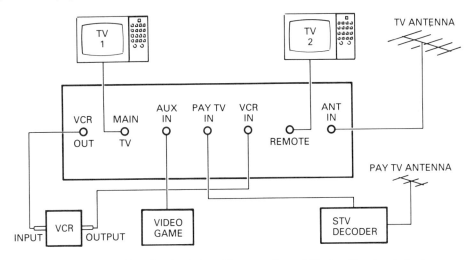

Figure 1.18. Video signal switcher. (Courtesy Channel Master, Div. Avnet, Inc.)

An arrangement of this kind does not follow the usual horizontal and/or vertical setup so common to block diagrams of components. In this case the blocks can be arranged in any convenient manner, but even here it is advisable to draw the lines so that as few of them cross as possible. Each of the lines represents a two-wire coaxial cable ending in coaxial type plugs.

Connector block diagrams are also useful when setting up an in-home high-fidelity system (Figure 1.19). The wiring behind such a system often looks like a maze and can become confusing. The connector block diagram is useful in helping to prevent connecting errors.

Block diagrams of this kind supply no information about the kinds of cables or connectors to use, although this information could be supplied in a separate footnote or parts list. In this illustration all of the lines represent coaxial cables with the exception of those going to the speakers. Note, in this illustration, that the components are arranged so that none of the connecting lines cross each other.

BLOCK DIAGRAMS FOR SIGNAL TRACING

If a block diagram of a component is available it can be of considerable help in servicing. Not only does it supply an overall view, but it can help servicing proceed in a logical and straightforward manner. A test signal can be applied to the input ter-

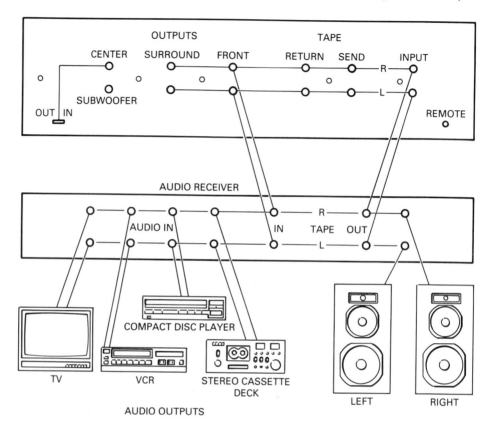

Figure 1.19. Connections for in-home high-fidelity system.

minals (Figure 1.20), and an indicating instrument such as a voltmeter or oscillo-scope across the output. The test signal is applied to stage after stage until the one that will not pass the signal is located. Alternatively, the test signal can be injected in the final stage and then worked backward to the input. In some instances the com-ponent will give recognizable symptoms so that the suspected stage can be checked immediately without going through this entire process.

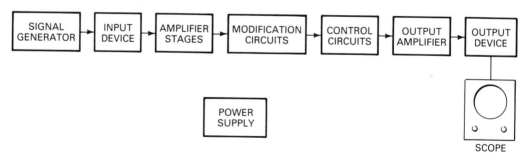

Figure 1.20. Use of block diagram for signal tracing.

It isn't always necessary to have a test indicator across the output since the component's speakers or picture tube may serve. It may also be necessary to use a test probe such as a demodulator type for checking the output of radio and intermediate frequency stages.

Quite often a circuit diagram will be available for the component to be repaired. While such a diagram is extremely helpful and in some instances absolutely essential, the block diagram can expedite the work. If no block diagram is available, the one in Figure 1.17 can be used for radio receivers or high-fidelity systems. The block diagram shown later in Figure 6.6 in Chapter 6 can be used for TV receivers.

BLOCK DIAGRAM FLOWCHARTING

A flowcharting worksheet (Figure 1.21) is ordinarily used preparatory to writing a computer program, but it can also work as an aid in servicing receivers, video cassette recorders, television sets, and other electronic components. The flowcharting worksheet is a preprinted form, and when it is inscribed with symbols it is called a "flowchart" and is then a member of the block diagram family.

The flowcharting worksheet consists of five columns and ten rows of blocks in dashed line form. Each column is identified by numbers, and so the first is the one column, the second is the two column, and so on. Each row is identified by letters with the first row called the A, the second the B, with the final row at the bottom of the form the K row. With this alphanumeric arrangement, any block on the page can be located quickly.

The Flowcharting Template

Unlike an ordinary block diagram, the symbols used in a flowcharting worksheet have specific meanings. While these symbols were originally designed to be used for computer programming, some have been selected as appropriate for electronics and are now being used as a servicing guide.

In a block diagram, the rectangles or squares can be drawn freehand or with the help of a template, but even with the limited number of geometric forms used in a flowcharting worksheet, a template, shown earlier in Figure 1.6, is very helpful. The rectangular outline shown as item 23 in the drawing of the template can be used for drawing block diagrams other than flowcharts.

Flowcharting Symbols

Although the flowcharting template shows a large number of symbol shapes, just a few of these are used for electronics. These include the decision symbol (No. 15 in the template), the processing symbol (No. 23), the terminal symbol (No. 7), and the connector symbol (No. 14).

Figure 1.21.   Flowcharting worksheet.

Terminal symbol

One of the advantages of the flowcharting block diagram is that there is no doubt about the start or finish since the terminal symbol is always used for both actions. Start and finish are comparable to the input and output of the other types of block diagrams.

Decision symbol

The decision symbol is a go, no go, or a yes or no symbol. If one apex of the symbol, usually at the right, is yes, then the apex at the bottom of the symbol is no. These yes/no positions can be transposed, and so it is possible to have yes/no marked on any apex. Usually the apex is at the left or right, or at the bottom. But no matter where they are positioned, the words yes/no must accompany the symbol.

Connector symbol

In some instances a number of columns in the flowchart may be required (Figure 1.22). In that case a column will not end in a terminal symbol since this would indicate the end or output. Instead, a connector symbol is used and the action continues at the start of the adjacent column. Movement is always from the top to the bottom of one column, then continuing at the top of the adjacent column to the right. Thus, if the flowchart starts and ends in column 1, the next usable column is 2, beginning with box A2.

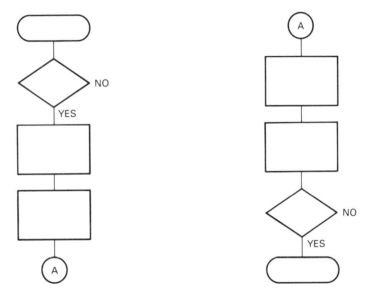

Figure 1.22. Use of a connector. The movement or flow is from A on the bottom left to A at the top right.

Usually, a flowchart starts with the first column, but this is not an absolute requirement—any column can be used. Nor is it necessary to complete an entire column before moving on to the next (Figure 1.23). In this flowchart each symbol was drawn using a template inside the dashed-line blocks on the preprinted flowcharting worksheet.

Processing symbol

The processing symbol is used for any action or work to be done. Typical phrases to be entered in this symbol would be "check audio input," "replace IC," "measure input signal voltage," and so on.

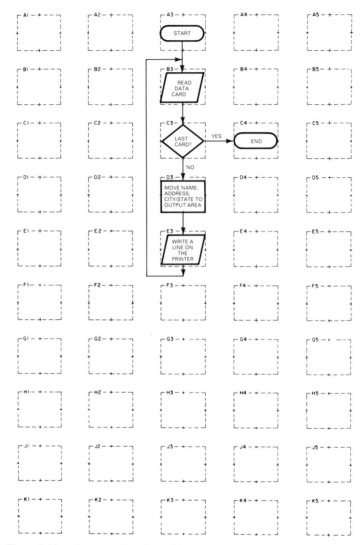

Figure 1.23. Computer flowchart. (Courtesy Programming & Systems, Inc.)

FLOWLINES

As in the case of other types of block diagrams, each symbol is connected to preceding and following blocks by solid lines called "flowlines." To avoid any misunderstanding, each flowline should have an arrowhead to indicate the direction of the work, although this procedure is commonly ignored.

USING THE FLOWCHART

It is easy to acquire experience in flowcharting. All that is needed is a template and a pad of preprinted flowcharting forms. Any activity, such as buying a newspaper, reading a book, commuting, or making a phone call (Figure 1.24) can be flowcharted.

The flowcharting technique can be applied to electronics servicing problems. There are two possible types. One is a flowchart covering an entire component, starting at the input and ending at the output. The purpose of such a flowchart would be to localize the fault to a particular circuit. Following this, a flowchart could be prepared to narrow the search to a specific part.

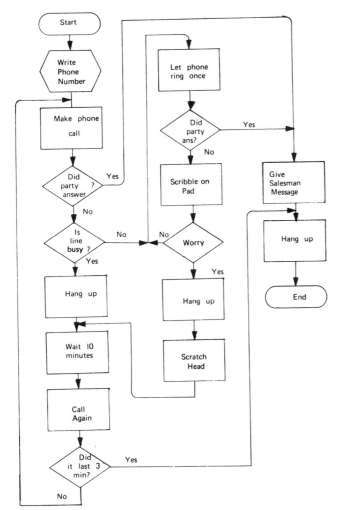

Figure 1.24. Flowchart for daily activities. (Courtesy Programming & Systems, Inc.)

Figure 1.25 illustrates a flowchart used for checking the operation of an RF circuit. The flowchart can be as detailed or as general as wished. It would be helpful to maintain a file of flowcharts used in servicing, including any corrections or changes made over a period of time. The flowcharts could then be used on a continuing basis.

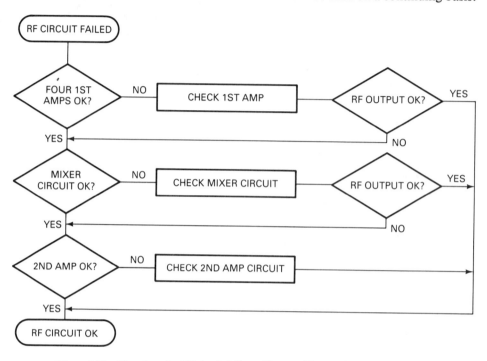

Figure 1.25.   Flowchart for RF circuit failure. (Courtesy Kyocera International, Inc.).

Servicing flowcharts are not only used by independent servicing technicians, but also in the servicing departments of manufacturers. The charts are based on the experience of those employed in these departments, are corrected and updated, and are then published by the manufacturer (Figure 1.26) for use by their servicing personnel.

ADVANTAGE OF FLOWCHART SERVICING

The servicing flowchart is desirable for several reasons. By following the logical path or paths indicated by the flowchart, tests that were made will not be repeated. If the flowchart is well made, no servicing steps will be accidentally omitted. Also, it does not follow that the flowchart must be slavishly followed right from the initial step. It may be that some of the earlier steps could be omitted. Finally, the flowchart lends itself to a written record of accumulated experience. With such experience, flowcharts can be updated with the diagram a constant reminder as to how the testing

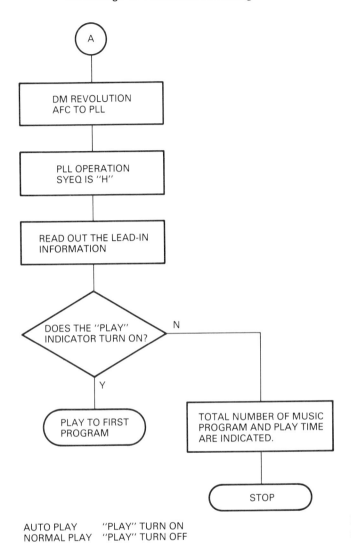

Figure 1.26. Flowchart for service technicians not factory associated.

AUTO PLAY	"PLAY" TURN ON
NORMAL PLAY	"PLAY" TURN OFF

should proceed. If electronics testing apprentices are to be trained, a complete set of servicing flowcharts, based on prior practical experience, form an excellent guide. In some instances servicing flowcharts are kept for specific models of components, with notes added concerning faults common to these models.

A block diagram showing the movement of a signal from the input to the output and servicing flowcharts do not compete, because both are essential. Block diagrams are useful for getting an overall understanding of how a group of related circuits work together; the flowchart shows how to find a fault in a specific circuit or circuits.

SHADOWED BLOCK DIAGRAM

Some diagrams have their blocks accentuated by a dark shadow outlining one or two sides of each block (Figure 1.27). It does not contribute to the detail or accuracy of the diagram, but some users may feel it improves its appearance. In some diagrams reversing Figure 1.28A or cross-hatching a block (Figure 1.28B) is done to call attention to a specific circuit.

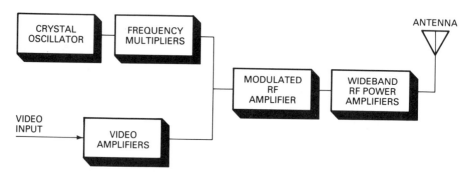

Figure 1.27.   Shadowed block diagram.

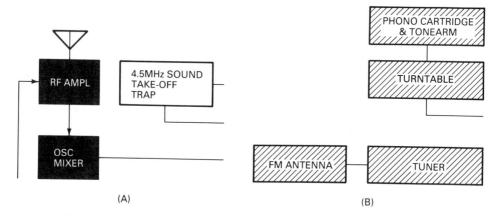

Figure 1.28.   Reversing (A) is sometimes used in block diagrams, or cross-hatching (B).

IC BLOCK DIAGRAM

An integrated circuit (IC), described in more detail in Chapter 5, consists of active elements (such as transistors) and passive elements (such as resistors) mounted on a substrate made of a semiconductor material. These parts form one or more circuits, with the entire IC contained in an insulating package. The IC is equipped with numbered terminals (Figure 1.29) to which connections can be made.

The IC is usually identified by a manufacturer's part number printed on the package. One or more ICs may be used in a component. Since the IC contains a cir-

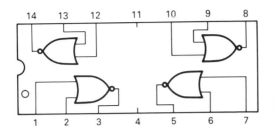

Figure 1.29. Diagram of an integrated circuit.

cuit, or circuits, these may be supplied separately. The outline for an IC is usually a solid line, but sometimes a dashed line is used to emphasize the fact that the IC is a packaged circuit.

LOGIC GATES

A logic gate is an electronic switch, and these are detailed in Chapter 7. Each type of gate has its own specific symbol. Gates are manufactured as ICs with one or more gates mounted on a substrate. Unlike the blocks used in a block diagram, the symbols for gates convey information as to the gate function. The symbols for ICs and gates may be combined with blocks in a block diagram. When gates are shown within the outline of an IC, what we actually have is a block within a block. In some instances a block may contain a circuit diagram, either a subcircuit (Chapter 4) or a partial circuit (Chapter 6).

MULTIPLE OUTPUTS

A component such as a TV monitor may be equipped for multiple inputs and a TV receiver for multiple outputs. A high-fidelity system could be able to accommodate various inputs and outputs. For multiple outputs these are usually arranged vertically in a block diagram (Figure 1.30). The outputs may be in parallel, supplied by a signal from a single source, or they may be independent, with each output wired to its own source voltage.

CIRCUIT DIAGRAM BLOCKS

Sometimes, in a circuit diagram blocks will be put around various components or partial or complete circuits. This may be done to distinguish one circuit from another. Except in certain cases, such as the squares or rectangles drawn around speakers, these blocks do not represent enclosures.

In some instances dashed lines (Figure 1.31A) are used to emphasize the relationship of one group of blocks to another even though these may not be adjacent in a block diagram. In this drawing the four blocks at the left are electronically related to the four blocks at the right while the three blocks in the center represent an in-

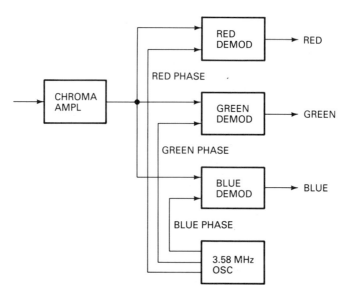

Figure 1.30. Block diagram in vertical form.

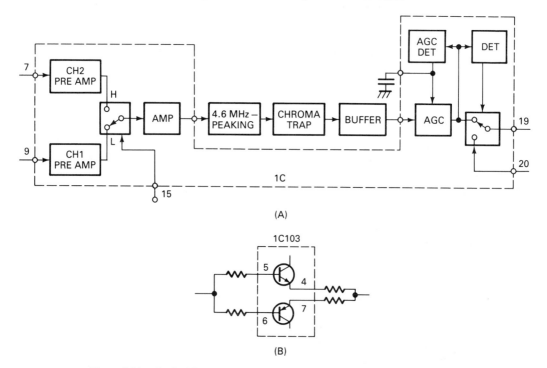

(A)

(B)

Figure 1.31. Dashed lines can be used to represent physical closeness of certain parts in an IC (A) or a complete IC (B).

termediate step. In Figure 1.31B, a dashed line is put around a pair of symbols to indicate that the two parts function as a unit.

SIGNAL DIVISION OR SEPARATION

A block diagram can be used to indicate signal division or signal separation (Figure 1.32). In an arrangement of this kind it is advisable to equip the lines that connect components with an arrowhead to show the movement of the signal.

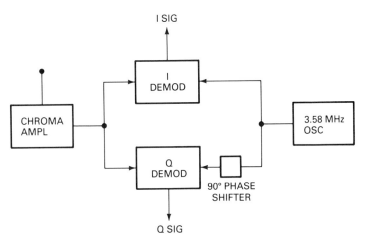

Figure 1.32. Blocks can be used to indicate signal separation.

SEMICONDUCTOR BLOCK DIAGRAMS

Sometimes, to explain the functioning of a part, it may be shown in block form (Figure 1.33). In this case the part under examination is a P-N-P transistor combined with the electronic symbols for batteries. It may be easier to visualize the direction of flow of the currents through the transistor when it is arranged in this form.

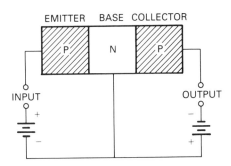

Figure 1.33. Block diagram form of a P-N-P transistor.

SYSTEM BLOCK DIAGRAMS

Block diagrams can be used for supplying an overall view of a complete component, such as a receiver, amplifier, compact disc player, and so on, but it is also useful for supplying information about the operation of a TV station and TV receivers. Figure 1.34 shows the overall functioning of a microwave relay system. In this system the signals are beamed from the transmitter to a specific receiving location.

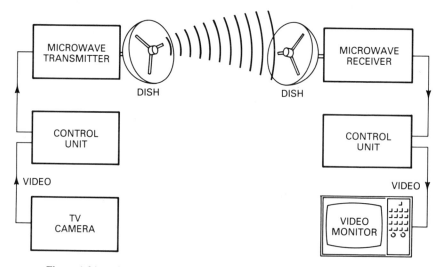

Figure 1.34. Block diagram used to supply overview of a complete system.

This diagram, and others that use blocks, seldom supply detailed information about the components they use. For more specific information, each of the blocks shown in this drawing would have its own set of block diagrams. The advantage of the overall diagram is that it supplies a view that gives an immediate (although basic) understanding of this system and is much easier to visualize than a detailed, written explanation.

In some cases a system such as this one has a series of block diagrams, with each successive diagram more detailed than the one that precedes it. Thus, following the diagram in Figure 1.34 there could be a block diagram of the microwave transmitter, and then another diagram of each circuit in the transmitter, and finally a diagram of the details of any ICs or gates that are used. This could be followed by a series of block diagrams for the microwave control unit, and so on.

2

Pictorials and Symbols

As its name implies, a pictorial diagram (Figure 2.1) uses drawings or photos of electronic parts or a combination of these as a way of depicting an electronic circuit. Block diagrams are not only simpler for the electronics novice but are also of great advantage in illustrating sophisticated systems. The pictorial is also advantageous, because it can show exactly how connections are to be made. For the beginner it is extremely helpful since it encourages an easy acquaintance with electronics parts.

In some instances a pictorial may be combined with a block diagram. This may be done to indicate the relationship of the part shown in the pictorial diagram to other circuits. Pictorial diagrams may be used alone or preceded or followed by a block diagram. Pictorials are supplied with kit building projects, an excellent way of getting hands-on experience in electronics.

Aside from the difficulty of drawing pictorials or obtaining suitable photos, this method does have several serious disadvantages. In some instances the parts to be depicted are potted, or in some type of container, and so little can be learned from observing them. Also, the appearance of many electronics parts can be quite dissimilar. This means a pictorial for one part may not be applicable to another, even though they may perform the same function.

While a pictorial is usually that of a complete unit, it can also show a portion of an assembly, that is, just a few electronics parts that have been put together (Figure 2.2).

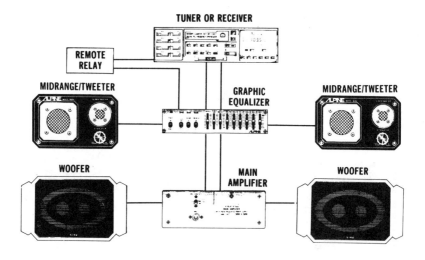

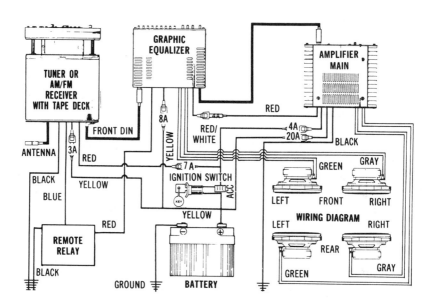

Figure 2.1. Two pictorials of the same system. Front view (A); wiring diagram (B). (Courtesy Alpine Electronics of America, Inc.)

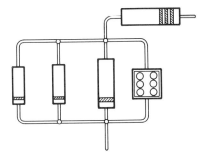

Figure 2.2. Partial pictorial.

WIRING DIAGRAMS

The pictorial diagram, sometimes called a wiring diagram, not only shows the positioning of electronics parts, but also has lines that represent the connecting wires. While these lines show how parts wiring should be made, they may or may not represent the actual positioning of these wires. In some instances the wiring diagram supplies data concerning the wires to be used.

CONSTRUCTION DIAGRAMS

Sometimes more than one pictorial is used in a kit building project. One of these is the wiring diagram, another is a construction diagram showing how the parts are to be mounted (Figure 2.3). These may supply dimensions for holes to be drilled, the position of the various controls, the location of brackets or other hardware, information concerning jacks, and so on.

Quite often a printed circuit board is used, especially in a kit. If the chassis and/or panel are prepunched, it then becomes a simple matter of identifying the various parts correctly and mounting them. If a printed circuit board is used, then most of the connections will be prewired and so very little actual wiring will be required. Any wiring that is used is called "point-to-point" to distinguish it from printed circuit wiring.

ASSEMBLY DIAGRAMS

An assembly diagram is a pictorial that shows in some detail how various parts are to be joined. In some instances more than one such diagram is required if the overall assembly is complex. But such a diagram is helpful even if it is for nothing more than mounting a bracket.

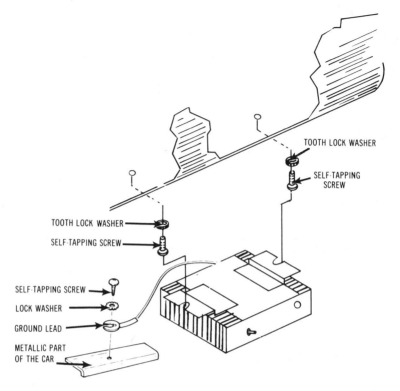

Figure 2.3. Pictorial detailing parts mounting. (Courtesy Alpine Electronics of America, Inc.)

DETAIL DIAGRAMS

A detail diagram is one that supplies the dimensions of a part that is to be mounted (Figure 2.4). In some instances a detail diagram will be associated with an assembly diagram, as a separate drawing, or the measurements may be made part of the assembly diagram. Dimensions are supplied in the English system (inches or feet) or the metric (millimeters or centimeters), or both.

PRINTED CIRCUIT DIAGRAMS

A printed circuit is a member of the family of pictorial diagrams. It is widely used and so is usually considered as being in a category of its own. Printed circuits, operational amplifiers (op amps) and integrated circuits (ICs) are closely associated and so receive separate and detailed treatment in Chapter 5.

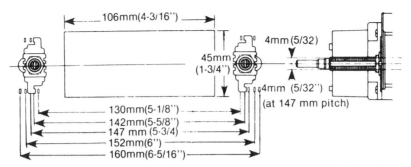

Figure 2.4. Detail diagram supplies parts dimensions. (Courtesy Alpine Electronics of America, Inc.)

COMBINATION PICTORIALS

In some instances two electronic components may be so closely associated that they are illustrated by a combined pictorial and a symbol that also emphasizes the combination. For example, a volume control and a switch may be mounted on a single shaft so that turning the switch to its "on" position also puts the volume control into its maximum resistance position. A coil may be in series or in parallel with a capacitor, and this will be shown in a pictorial and in combination form by a pair of symbols. The purpose here is to show that the individual parts are either mechanically or electrically joined. But while these parts may be associated physically they aren't necessarily near each other in the final circuit diagram. In the case of the switch/volume control combination, the switch would be part of the power supply, the volume control part of the audio amplifier.

WAVEFORM PICTORIALS

Sometimes, to indicate a particular circuit operating condition a voltage waveform may be included at the input or at the output, or both. Waveforms are pictorials and are often just idealized representations of the actual voltage. In some instances the peak-to-peak value of the wave is included. These waveforms may be included with either a pictorial or a circuit diagram and are commonly used in diagrams of TV receivers. Typical waveforms would include a sine wave, positive and negative pulse waveforms, trapezoidal waveforms, and so on (Figure 2.5). The waveforms may be drawings or photographs taken from the screen of an oscilloscope.

CALLOUT PICTORIALS

A callout pictorial consists of a view of the front of a component or its interior, such as a receiver or an amplifier or similar equipment, using various methods of identifying parts. A commonly used method is to draw lines to each and labeling them

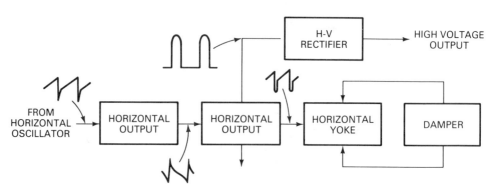

Figure 2.5.   Block diagram showing use of waveforms.

with letters, numbers (or a combination), or a descriptive word (Figure 2.6). Pictorials of this kind are advantageous in servicing since they permit the quick location of operating controls, when using a front view, and make it much easier to find individual parts when the pictorial is an underchassis view.

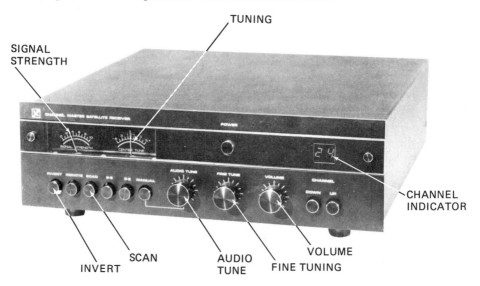

Figure 2.6.   Callout pictorial. (Courtesy Channel Master, Div. Avnet, Inc.)

INSTRUCTIONAL PICTORIALS

In some instances an instruction can be conveyed far more effectively by picturing it rather than describing it by written text. Thus, the action to be performed (Figure 2.7) is self-evident. When a number of steps are to be taken there can be a pictorial for each one and in some instances accompanied by written guidance. However, with the help of the pictorial this can be kept to a minimum.

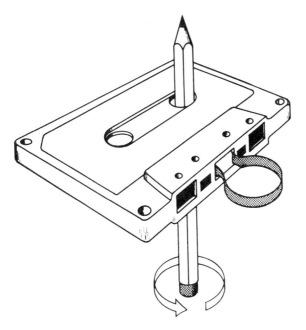

Figure 2.7. Instructional pictorial. (Courtesy Alpine Electronics of America, Inc.)

HARDWARE PICTORIALS

A surprising amount and variety of hardware is associated with electronics including machine screws, metal screws, self-tapping screws with hex heads, slotted heads, or Phillips heads (Figure 2.8).

There are also flat washers, split washers, external and internal star washers, and of course a large variety of knobs. When pictorials are supplied, identification of hardware becomes much easier.

Lugs

A lug is a type of connector and is commonly used for connecting a wire to a machine screw being used as a terminal. The antenna terminal board on a TV receiver may have a pair of machine screws for a connection to twin lead transmission line with that line equipped with a pair of spade lugs. There is no electronic symbol for the lugs since they are regarded as hardware. Still, the lug is a connector and is usually shown in pictorial form.

Clips

There are various kinds of clips used for making temporary connections, including alligator, crocodile, microtest, minitest, insulated and noninsulated, car battery, jumbo, claw, and others. These are always shown in pictorial form. However, there are electronic symbols for connectors such as plugs and jacks.

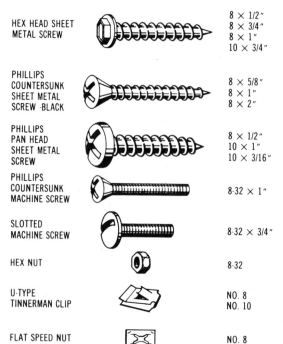

HEX HEAD SHEET METAL SCREW	8 × 1/2" 8 × 3/4" 8 × 1" 10 × 3/4"
PHILLIPS COUNTERSUNK SHEET METAL SCREW -BLACK	8 × 5/8" 8 × 1" 8 × 2"
PHILLIPS PAN HEAD SHEET METAL SCREW	8 × 1/2" 10 × 1" 10 × 3/16"
PHILLIPS COUNTERSUNK MACHINE SCREW	8-32 × 1"
SLOTTED MACHINE SCREW	8-32 × 3/4"
HEX NUT	8-32
U-TYPE TINNERMAN CLIP	NO. 8 NO. 10
FLAT SPEED NUT	NO. 8

Figure 2.8. Hardware pictorial. (Courtesy Alpine Electronics of America, Inc.)

EXPLODED VIEW PICTORIALS

An exploded view (Figure 2.9) shows all the parts of a component and their physical relationships. An exploded view is helpful when physical assembly is required, but this view does not show any required wiring. Diagrams of this kind often accompany kits. In some instances the component involved is a complete unit but requires installation or assembly of some kind. The one in this drawing is an exploded view pictorial of the under-dash details of an auto radio.

MASKED DIAGRAMS

In this diagram (Figure 2.10) all of the radio parts except those being described are masked. This makes it easy to find certain parts without being confused or distracted by a large number that are nearby.

CUTAWAY PICTORIALS

Electronic parts are sometimes encapsulated, potted, or enclosed in some kind of sealed container preventing examination of interior parts or construction, such as integrated circuits, transformers, chokes, inductors, and so on. With a cutaway pictorial (Figure 2.11), some part of the component is removed to enable the interior to become visible.

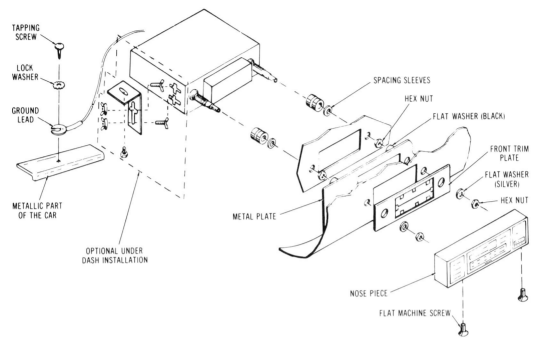

Figure 2.9. Exploded view pictorial of auto radio installation. (Courtesy Alpine Electronics of America, Inc.)

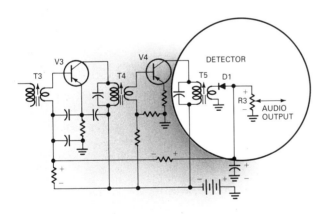

Figure 2.10. Masked diagram emphasizes circuit being analyzed.

ELECTRONIC SYMBOLS

To anyone not familiar with electronic symbols, a first view of a complete circuit diagram can be discouraging. The mass of symbols and their connecting lines can be incomprehensible. For this reason it is preferable to start with a study of individual symbols, but there are so many of these that this is also discouraging.

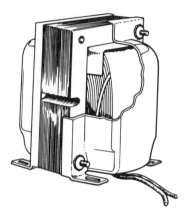

Figure 2.11. Cutaway pictorial permits examination of interior of potted or encapsulated components.

Still, there are two hopeful approaches to an understanding of complex circuit diagrams. The first is that every diagram, no matter how extensive, consists of the same few symbols repeated over and over again. And the second is that there is an easy approach to becoming familiar with electronic symbols by using pictorials.

SYMBOL STANDARDIZATION

There are standards for symbols contained in USA Standard Graphic Symbols for Electrical and Electronics Diagrams (ANSI Standard Y32.2d-1970) made available by the American National Standards Institute, Inc. However, this does not mean that those who draw diagrams for manufacturers, book or magazine publishers, distributors of technical literature, technical institutes, schools and colleges follow the symbols precisely. Many set up their own standards, and while the symbols may be consistent within their own organization, they may well be at variance with the symbols used by others.

There is still another problem associated with standardization and that is that electronics is a fast-moving science. Components are constantly being developed, with new symbols being produced just as rapidly. And, in some instances, older products are discarded.

Electronic symbols do not exist in a world of their own. Any component using AC power may require industrial electronics symbols. There are also components that require motors such as turntables and compact disc players, and, strictly speaking, these motors call for industrial or electrical symbols. For these reasons, such symbols have been included in this chapter.

Symbols are electronic shorthand. Circuit diagrams are composed of these symbols with the symbols connected by lines to indicate their relationship. They are specified as electronic symbols to distinguish them from symbols used in electrical or industrial work.

TYPES OF SYMBOLS

The word "symbol" is used to represent those that are used in circuit diagrams as well as abbreviations and acronyms. There are four types of symbols: graphic, numeric, alphabetic, and alphanumeric (Figure 2.12). Of these the graphic is the most important when working with diagrams, but the others may be used to furnish supplementary information.

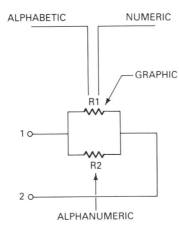

Figure 2.12. Graphic symbols and methods of identification.

Graphic Symbols

The graphic symbol is used to represent electronic components in pictorial style. Graphic symbols are easy to draw, either freehand or with the help of a template.

Numeric Symbols

Numeric symbols use digits and are helpful in identifying a specific part of a circuit diagram. The numbers 1 and 2 in Figure 2.12 indicate the input terminals. The small circles are the graphic symbols, and so what we have here is a combination of graphic and numeric.

Alphabetic Symbols

An alphabetic symbol uses one or more letters of the alphabet. The letter "T" (or "Tr") may be associated with the graphic symbol for a transformer. The abbreviation "spkr" is sometimes found together with the graphic symbol for a speaker.

Alphanumeric Symbols

An alphanumeric symbol is one that consists of one or more letters in association with one or more digits. C1 and C2 are alphanumeric symbols used to indicate a pair of capacitors.

ABBREVIATIONS

A variety of abbreviations may be used in conjunction with graphic symbols. There is no standardization. An abbreviation for a transformer, for example, could be the letter "T," the abbreviation "Tr," or "Xformer." The abbreviation often includes the first letter (or letters) of the word describing a component. Thus, "ant" is used for antenna, "gnd" for ground, "IF" or "I.F." for intermediate frequency, "uhf" for ultra high frequency, and so on.

ACRONYMS

An acronym is a coined word and consists of the first letter, or first few letters, of a phrase describing some electronic component. Radar, for example, is derived from "*ra*dio *d*etection *a*nd *r*anging."

SYMBOL OBSOLESCENCE

In time, some electronics components become obsolete, and as they do so their graphic symbols also fade away. A number of these antiquated symbols are shown near the end of this chapter as an aid for those who are interested in or who are collectors of old-time components. In some instances the symbol is carried along but is changed in some way to indicate component modification.

ANTENNA SYMBOLS

The antenna (ant) and ground (gnd) symbols are two of the most important in circuit diagrams of all kinds. The antenna symbol by itself or in conjunction with the ground symbol often represents the input points of a circuit. These starting points are helpful when tracing the path of signal movement.

The most commonly used graphic symbol for an antenna consists of an inverted triangle with a vertical line bisecting this geometric figure (Figure 2.13A). This isn't the only antenna symbol; the others in this illustration have also been used.

It isn't possible to determine from the symbol much information about the antenna or its downlead, a transmission line connecting the antenna to the input of the receiver. However, the symbols in Figures 2.13A and B are commonly used to represent an antenna for AM reception. Another antenna symbol, Figure 2.13C shows two lines connected to the antenna, and these indicate two-wire transmission line (twin lead) and so the antenna, sometimes referred to as a balanced antenna, is an FM or TV type. Sometimes the same antenna is used for both FM and TV reception. For this antenna the downlead can be either twin lead or coaxial cable.

None of these symbols supply any data about the impedance of the antenna or of the downlead, although the twin lead in Figure 2.13C usually has an impedance of 300 ohms. Coaxial cable is rated at 75 ohms.

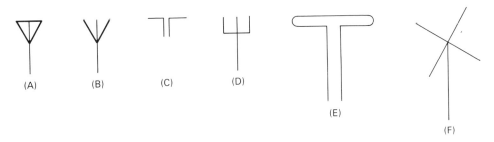

Figure 2.13.   The symbol in (A) is the one most commonly used to represent an antenna. Symbol (B) is used less often. (C) represents an FM or TV antenna; the symbol in (D) is antiquated. Symbol (E) is used for a folded dipole TV antenna while (F) is a Yagi.

Loopstick Antennas

In the early days of radio the receivers were quite insensitive, and so a long outdoor antenna was a requirement. But as tube gain improved, the antenna became smaller and finally became a separate component in the form of a large loop mounted on a wooden frame and positioned on top of the receiver (Figure 2.14A).

One of the advantages of the loop antenna was that it could be rotated on its vertical axis and so, when broadside to the transmitting station, was in a position to pick up a signal of maximum strength. It also acted as a crude tuning device, improving selectivity. A typical loop antenna consisted of 12 turns of insulated No. 18 wire, wound in a square, two feet on a side, the turns being separated one-half inch.

Eventually, even this indoor loop antenna became smaller, disappeared into the receiver, and was called a "loopstick," a reminder that a loop was still being used but was now mounted on something resembling a stick. It consists of a coil of wire wound on a small form containing a polyiron (ferrite) core with that core represented by dashed lines (Figure 2.14B). The remainder of the symbol is that of a coil (inductor). Dashed lines are used for the core to indicate it is made of powdered, not solid, iron. At one time loopsticks were tunable, indicated by an arrow in Figure 2.14C. The adjustment was often made to favor a weak station or one that was preferred. Receiver selectivity has improved to the point where loopstick tuning is no

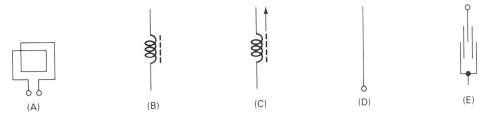

Figure 2.14.   Symbol for a loop antenna (A); nontunable loopstick (B); tunable loopstick (C); single element auto whip antenna (D) and telescoping antenna (E).

longer essential. The word "antenna" is abbreviated as "ant" and is sometimes used alongside the symbol.

There are two basic types of antennas used for auto radios. One of these is a nonadjustable type called a "whip" (Figure 2.14D). The other is a telescoping antenna often consisting of three (or more) vertical sections that fit into each other and so the antenna height can be adjusted (Figure 2.14E). Often though, the symbol for the whip is used for both types.

Antennas for TV sets are indoor or outdoor types. The outdoor antennas can be dedicated; that is, separate antennas are used for VHF and UHF signals, with another for FM, or they can be three-in-one for all TV and FM broadcasts. Indoor antennas can be built-in or add-ons. The VHF antenna is a telescoping rod, rotatable 360°, while the UHF is a circular type capable of being moved back and forth.

Antennas are passive devices; that is, they do not supply signal gain. When the word "gain" is used in connection with antennas, this is simply a comparison with a reference standard. An antenna that has gain is simply one that supplies more signal output than a reference standard working under similar conditions.

In some instances, though, a preamplifier is connected directly at the antenna output terminals, and so this combination antenna and preamplifier works as a combined system, referred to as "active." Technically, though, the antenna is still passive, because it is the preamplifier that is active.

GROUND

Ground (gnd) isn't a physical component in the same sense as resistors, capacitors, and coils. Nevertheless, one or more ground symbols are shown in circuit diagrams. Components such as receivers and amplifiers have a ground terminal positioned on the rear apron. Turntables even supply a separate wire leading from inside the component to be connected to ground. Ground is often considered as a voltage reference point with measurements taken with respect to ground. The abbreviation "gnd" sometimes accompanies the ground symbol. The word "ground" isn't accepted internationally—in Great Britain it is referred to as "earth."

The ground symbol (Figure 2.15A) consists of a series of parallel, horizontal lines with the longest at the top. At one time a ground symbol meant exactly that, a connection to earth or ground. Today, it more often means a common connection and as such is used to indicate a common line or tie point. Like the antenna symbol, that used for ground has a number of variations, but the one shown in Figure 2.15A is used most often. Figure 2.15B is an alternate symbol.

Ground is also a voltage reference. Thus, a potential of ten volts requires two points between which this voltage is measured. One of these is ground. A negative

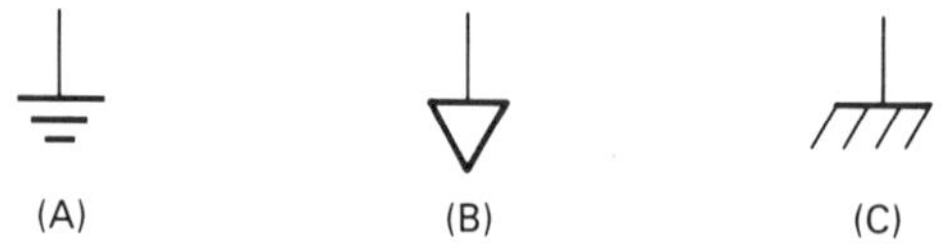

(A)   (B)   (C)

Figure 2.15. Preferred ground symbol (A); alternate symbol (B); floating ground (C).

potential, also measured with respect to ground, always uses a minus sign.

A ground symbol is a way of simplifying circuit diagrams. Those in Figures 2.16A and B are the same except that the one in B uses a ground symbol, making the circuit easier to read.

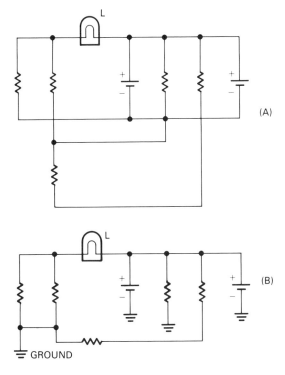

Figure 2.16. Circuits A and B are identical. Use of ground symbol (B) simplifies circuit.

Floating Ground

One side of the AC power line is grounded and is the same ground used in various components such as receivers, amplifiers, and transcription turntables. In these, either a metal chassis or common wire is used as the ground return. If this type of ground is not wanted, a conductor, such as a suitable length of wire, is used, to which all ground returns are connected. Known as a "floating ground," it is isolated from the chassis and the ground of the AC power line. The symbol for this type of ground appears in Figure 2.15C to emphasize its difference from a chassis or earth ground.

Mobile Ground

The ground for a mobile installation, such as that in a vehicle (car, truck, van), plane, or boat, can be the metal frame. If a component such as a radio receiver or a speaker has one of its terminals attached to the frame of the vehicle it is said to be directly grounded. However, some, such as speakers, use a floating ground, and instead of using the car frame as a conductor, require a pair of wires for connection to the output of a main or power amplifier. The advantage of a floating ground in any kind of vehicle is that it is superior to a direct ground. The frame of a vehicle isn't necessarily continuous, and its metal parts may rub against each other, producing noise in the output.

Most vehicles use a negative ground; that is, the frame is directly wired to the minus terminal of the battery. Some cars, though, particularly foreign makes, have a positive ground.

BATTERIES

A battery consists of two or more cells, connected either in series or parallel, but the word "cell" is gradually being phased out, and even a single cell is now called a "battery."

A battery is a DC voltage source. Its basic symbol consists of two parallel lines of unequal lengths, drawn either vertically or horizontally. Ordinarily a plus symbol (+) is positioned adjacent to the longer of the two lines and represents the plus terminal. Sometimes this symbol is omitted, but the longer line is understood to be plus (Figure 2.17A).

A battery can be represented by any one of a number of symbols (Figure 2.17B). If several batteries are used in a circuit they are sometimes identified as E1, E2, or B1, B2, and so on. The voltage supplied by the battery may be marked alongside the symbol. The batteries in Figure 2.17B are series connected; those in Figure 2.17C are in parallel.

The symbol itself does not supply any information about the type of battery (whether it is a dry cell or a wet storage type), its ampere-hour rating, or its physical

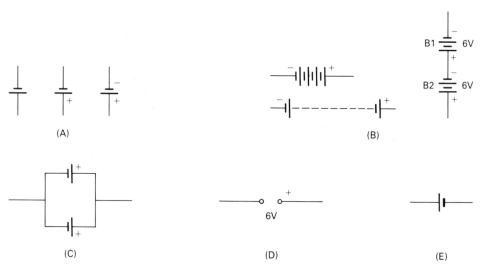

Figure 2.17. Battery symbols.

charactcristics. Sometimes two small circles are used to represent a battery, but since these are not recognized as a battery symbol, a voltage rating is included (Figure 2.17D). In some older circuit diagrams, the battery symbol consisted of two parallel lines, but with one of these representing the negative terminal thicker than the other (Figure 2.17E). The purpose here was to eliminate the need for using a minus sign to indicate polarity.

Current flow from a battery is generally understood to start at the negative terminal, move through some external circuit, and then back to the plus terminal. This is just a convenient fiction, because current flow is like a wheel—all its parts move simultaneously.

RESISTORS

Resistors represent one of the most important of all electronic components and appear, in numbers, in every type of equipment. The variety of resistors is impressive, and yet they can be represented by just a few symbols.

Resistors can be categorized in different ways, but one of the more commonly used is based on design. Resistors can be fixed, single variable, ganged (that is, two or more variable resistors mounted on a common, rotatable shaft), tapped, low or high wattage, light operated, heat operated, and so on.

Fixed Resistors

Figure 2.18A shows the basic symbol for a resistor, a fixed type in the sense that its value isn't changeable. In some instances the value of resistance, and possibly its wattage rating as well, are listed alongside this component.

Tapped Resistors

There are two types of tapped resistors, one in which the tap can be made to slide along the resistor permitting the selection of a specific value of resistance, and one in which the taps are permanently fixed. With the fixed tap type (Figure 2.18B), the resistance values cannot be changed. The slide type (Figure 2.18C) does permit a selection of different amounts of resistance. The symbol for the slide resistor always includes an arrow to indicate movement. Tapped resistors, whether fixed or adjustable, are wirewound and are designed to have a higher power dissipation rating than fixed

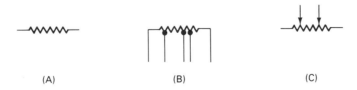

(A)   (B)   (C)

Figure 2.18.   Fixed resistor (A); fixed tapped resistor (B) and sliding tap resistor (C).

carbon type resistors. Tapped resistors are sometimes used as voltage dividers. The disadvantage of tapped resistors is that they are bulkier than the fixed, nontapped types.

Variable Resistors

Resistors can not only be fixed or tapped, they can be variable. Such resistors are continuously adjustable and can produce any desired value of resistance from close to zero to some predetermined value.

Individual variable units are the potentiometer (often abbreviated "pot"), resistors that carry very little current and work as signal voltage dividers, or as rheostats, designed to carry relatively large amounts of current. Potentiometers are used as volume controls, loudness controls; rheostats for controlling the current to some electronic part. A potentiometer may be made of a carbon-impregnated strip; a rheostat constructed of resistance wire.

Rheostats and potentiometers can be two terminal devices, but ordinarily have three. The same symbol (Figure 2.19A) is used for either potentiometers or rheostats.

Ganged Resistors

A pair of potentiometers can be mounted on a common shaft and rotated simultaneously. Known as an "attenuator," a typical unit consists of two variable resistors R1 and R2 (Figure 2.19B) designed to present a constant resistance to a source voltage (EMF or electromotive force). The attenuator can be adjusted so its input resistance equals the resistance of the voltage source. At the same time the attenuator drops the voltage to the amount required by the load. Electrically, the resistors are independent of each other. The dashed line shows that the two potentiometers are ganged.

In some instances a potentiometer is a one-time adjustment with the amount of resistance determined at the time of manufacture of the component using the variable resistance. The potentiometer is set to obtain a desired operating condition, a process known as "tweaking." The potentiometer uses a slotted head type

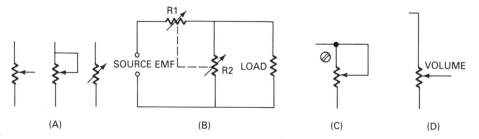

Figure 2.19. Symbols for potentiometers and rheostats (A); ganged variable resistors (B). Screw adjust potentiometer (C). Sometimes the screw symbol is omitted and the abbreviation "cal" (calibrate) is used instead or in addition to the screw symbol. In some circuits the function of the variable resistor is printed adjacent to the symbol (D).

shaft, adjustable by a screwdriver. Since the potentiometer is not a user control it is mounted on the rear apron and sometimes on the printed-circuit board inside the component. The symbol (Figure 2.19C) is that of a potentiometer with the drawing of a slotted screw head placed alongside it.

In some instances explanatory material is placed alongside the symbol, as in Figure 2.19D. This is optional, and there is no consistency, even within the same circuit diagram.

REPRESENTATION OF VALUES

The value of a component can be positioned above, below, or alongside it. There is no standardization from one circuit diagram to the next, although it is helpful if the same kind of value representation is used within a specific diagram. In Figure 2.20 there are four different ways of showing the value of a single resistor. While

Figure 2.20.   Four methods of indicating resistance values.

these could all be indicated in megohms, this would lead to the use of decimal fractions and is avoided when resistors are involved. As an example, a 5100-ohm resistor could also be designated as 0.0051 megohms, an awkward way of showing its value. However, while decimals are avoided when resistors are used, this does not apply to other components, such as capacitors, described later.

Prefixes for Resistors

The basic unit of resistance is the ohm, and because this unit is fairly small, alphabetic symbols are often required to avoid the use of unnecessarily large numbers. Whole numbers can be used for resistors having values of less than a thousand ohms, but above this amount it becomes convenient to use suitable symbols. The letter k is a symbol as a multiplier having a value of 1,000. A resistor rated at 5100 ohms can be written as 5.1k ohms, or simply as 5.1k. In some instances a capital letter is used instead, and the resistor value is designated as 5.1K. The letter m or M is a symbol as a multiplier having a value of one million. A 2m or 2M resistor is a two-million ohm resistor (2,000,000 ohms). The letter M is an abbreviation for megohm, often abbreviated as meg. A 1.2 meg resistor is a unit having a resistance of 1,200,000 ohms. There is no standardization in the use of these symbols, but it is helpful if there is symbol consistency within a single circuit diagram. There is no symbol or abbreviation for values of less than one ohm, and although such values do exist, they are rarely specified in a circuit diagram.

The capital Greek letter omega (Ω) or lower case omega (ω) are often used as an abbreviation for ohms. Thus, a 100-ohm resistor might be written as 100ω, and a 1,000-ohm resistor as 1Kω.

RESISTOR CODES

A code is a single letter, or a combination of letters, used to identify resistors in a circuit diagram, and for fixed, tapped, or variable resistors is the letter R. If a number of resistors are used they are followed by numbers, such as R1, R2, R3, and so on. In some cases both codes and values appear in a diagram. Ordinarily codes begin at the left and increase numerically moving across a diagram to the right. Thus, R1 would be the leftmost resistor, followed by R2, R3 and so on.

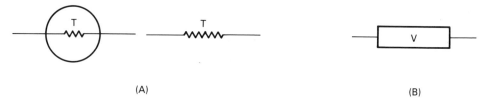

(A)   (B)

Figure 2.21. Symbols for thermistors (A) and a varistor (B).

Thermistors

Ideally, resistors should maintain a constant resistance for the conditions under which they work. Actually, resistors are sensitive to changes in temperature but can be constructed to have low-temperature coefficients of resistance; that is, they will maintain a reasonably constant value of resistance for substantial changes in temperature.

There are some applications, though, in which it is desirable to have the resistor sensitive to temperature. Specially made resistors with this idea in mind and made of a semiconductor material are known as "thermistors." The word "thermistor" is the abbreviation and union of two words—*therm*al and res*istor*.

Figure 2.21A shows possible symbols for thermistors. In some instances, the circle drawn around the resistive element is omitted. Usually the letter T is written above the resistor symbol to distinguish the thermistor from ordinary resistors.

Thermistors are made in two basic types, the two-terminal directly heated and the four terminal indirectly heated.

Varistors

Varistors (Figure 2.21B) are two-terminal voltage-sensitive resistors. Like thermistors (but unlike resistors) they use a semiconductor material (silicon carbide or carborundum) as the resistive element. The word "varistor" is an acronym coined from *vari*able res*istor*. Also known as voltage dependent resistors, they are unlike fixed resistors since their ohmic value is variable and is based on the voltage across them. Varistors are made in a variety of shapes including discs, rods, and washers and may (like resistors) be equipped with leads, but may also be supplied without them.

Photoresistors

A photoresistor is a resistor whose ohmic value varies depending on the quantity of light reaching it. Typically, a photoresistor has its maximum resistance when in complete darkness, with this resistance decreasing as the unit becomes more and more exposed to light. However, there is a limit to the drop in resistance, and the photoresistor never reaches a condition of zero ohms.

CAPACITORS

Next to resistors, capacitors are probably the most widely used of all electronic components and are available in a wide variety of sizes and capacitance.

Fixed Capacitors

Fixed capacitors can be axial types in which a connecting lead is brought out from the center at each end, or radial, in which both leads are brought out from one of the ends. Leads for capacitors can extend horizontally or vertically. These include polarized electrolytic, nonpolarized electrolytic, disc ceramic, mica, mylar, paper, polystyrene, tantalum, and metal film types.

A variety of symbols (Figure 2.22) are available, but the most widely used is that in Figure 2.22A and consists of a pair of lines one of which is straight, the other slightly curved. At one time, and still appearing in some old circuits, both lines are straight (B). Another type, once quite popular, is the feedthrough (C) with the capacitor made to force-fit in a chassis hole, or soldered into position in that hole. It was a way of bringing connections for a capacitor from below a chassis to some point above it.

The electrolytic capacitor may be polarized, in which case one of the lines of the capacitor is marked with a plus (+) symbol (Figure 2.22D). The nonpolarized electrolytic does not use this symbol. In some instances several electrolytic capacitors are enclosed in a common container. The negative terminals of these capacitors are interconnected internally, while the positive terminals are independent (Figure 2.22E).

(A)   (B)   (C)   (D)   (E)

Figure 2.22. Symbols for fixed capacitors. (A) Most widely used symbol; (B) symbol found in some older circuit diagrams; (C) feedthrough; (D) polarized electrolytic; (E) multi-electrolytic.

Variable Capacitors

Variable capacitors are those in which the capacitance can be adjusted from some predetermined minimum to some maximum value. Various symbols are used for these capacitors, but the one that is preferable is that shown in Figure 2.23A. It resembles

Figure 2.23. Variable capacitors. That in (A) is widely used; the symbol in (B) is no longer popular. (C) is a split-stator variable while (D) is a ganged type.

the symbol for the fixed capacitor except that the curved part of the symbol terminates in an arrow. Another symbol (Figure 2.23B) once more widely used, has an arrow drawn through it.

Variable capacitors can be single units or two or more mounted on and rotated by a common shaft (Figure 2.23C and D). The capacitors are electrically independent of each other and require their own connections. Also known as a ganged variable, it can consist of as few as two, while some have had as many as six capacitors. The dashed line indicates that the capacitors are ganged.

Capacitance Values

Unlike resistors, decimal values are used for capacitors. The reason for this is that the basic unit of capacitance is the farad, an extremely large amount. Consequently, capacitance values can be listed in either fractional or decimal form, but with the use of suitable prefixes, whole numbers can be used as well.

All capacitance units are expressed in terms of submultiples of the farad and include the microfarad (μF) or millionth of a farad, and the picofarad (pF) or millionth of a millionth of a farad. The picofarad was known at one time as the micromicrofarad ($\mu\mu$F) and is no longer being used, but you may still find it on some older circuit diagrams.

COILS

A coil, also known as an inductor, consists of one or more turns of wire, usually wound in circular form, although other geometric shapes are sometimes used. The coil may be wound on a form made of impregnated paper or plastic, or can be self-supporting, but in either case is known as an "air-core" type. In some instances a metallic core is used, and so the coil is referred to as an "iron-core" type. There are two variations of iron cores, powdered and solid. Powdered iron cores are often adjustable, while the solid iron-core type is fixed in position. The purpose of the adjustable core is to permit the selection of a wanted value of inductance.

Figure 2.24 shows various symbols used for inductors. For an air-core coil, that in Figure 2.24A is most commonly used. The dashed lines around the coil in B represents an enclosure. Generally made of metal, it protects the coil against induced voltages due to stray magnetic fields. The shield is often grounded, but no ground symbol is used to indicate this. Powdered iron cores (C) are represented by one or two vertical dashed lines, with these terminating in arrows if the core is an adjustable type. Solid iron cores (D) are indicated by unbroken lines, and since these cores are

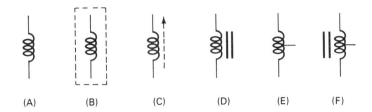

Figure 2.24. Symbols for inductors. Air-core coil (A); shielded air-core (B); tunable polyiron core coil (C); laminated (fixed) iron-core coil (D); tapped air-core coil (E); tapped iron-core coil (F).

fixed in position no arrows are used. Like a resistor, a coil can be a fixed type or may have one or more taps (E and F).

As in the case of capacitors and resistors, coils have a large variety of uses. They work as antenna coils (connected between antenna and ground), as a part of tuned circuits, as feedback coils in oscillators, as IF (intermediate frequency) transformers, power transformers, audio transformers, voice coils, in relays, in tuned circuits, and so on.

Magnetic Field Symbol

Whenever an electrical current flows through a coil it produces a magnetic field around the coil, and the coil becomes an electromagnet. This magnetic field can be represented by elliptical dashed lines around the coil, but the lines are not generally shown unless there is some special reason for doing so.

Sometimes the electromagnetic field around a coil may be strong enough to reach over and surround other coils, interfering with their work. To prevent this the coil may be surrounded by a metallic shield, which is shown by dashed lines. A ground symbol may be shown connected to the shield symbol, but quite often this symbol is omitted.

TRANSFORMERS

A transformer (Figure 2.25) consists of one or more coils all having a special relationship. Fundamentally a transformer uses two coils, a primary winding and a secondary winding, and like other coils can be air core or iron core types.

The number of turns of wire of the primary winding compared to the number of secondary turns is one way of identifying the transformer. If both primary and secondary windings have the same number of turns, the transformer is called a 1:1 (one to one) (A). If the secondary has more turns than the primary, the transformer is a step-up unit (B); if the secondary has fewer turns than the primary it is a step-down type (C). The 1:1 transformer and the step-up and step-down units can be iron core (D) or air core units (E). The core can also be fixed or variable.

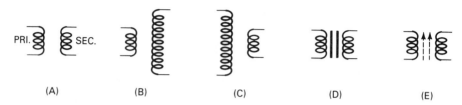

Figure 2.25. Transformer symbols. (A) 1:1; (B) step-up; (C) step-down; (D) fixed iron core; (E) variable iron core with both coils tunable. This symbol has a number of variations. One of the dashed lines can have a downward pointing arrow.

The primary of a transformer is usually connected to a voltage source. That voltage could be supplied by the power line or else it can be a signal voltage. The secondary of the transformer is connected to some device to which the voltage is to be delivered, without change, or stepped up or stepped down.

The Autotransformer

An autotransformer (Figure 2.26A) is a transformer having a single winding. It uses a tap so the single winding functions both as a primary and secondary. Figure 2.26A shows the autotransformer used as a 1:1 unit; while the symbol in B is that of a step-up unit, and in C a step-down. Note that the autotransformer has only three connections, while those in Figure 2.25 are four terminal devices.

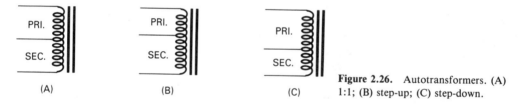

Figure 2.26. Autotransformers. (A) 1:1; (B) step-up; (C) step-down.

The Power Transformer

A power transformer is a multiple winding type; that is, it consists of a primary, which may or may not be tapped, and a number of secondary windings (Figure 2.27).

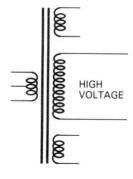

HIGH
VOLTAGE

Figure 2.27. Power transformer. Quite often it is a combined step-up and step-down type. The primary winnding may be tapped. The secondary may have more than the three winndings shown here.

The transformer in this illustration connects to the AC power line and delivers various values of AC voltage. It is, in effect, a combined step-up and step-down transformer. The wires of the transformer are color coded, with the insulation using this coding. However, there is no standardization and so the colors are not universally applicable.

The primary winding has three input leads, but only two of these are used, depending on the connecting instructions issued by the manufacturer. The winding marked "high voltage" is a step-up winding. The other windings are step-down types. The input voltage is AC, and so are the various voltages supplied by the secondaries. The unit uses a fixed iron core. The center connection of all the secondary windings is a tap to the electrical, not physical, center of the transformer winding. As shown by the solid vertical parallel lines separating the primary and the secondary windings, this is an iron-core type. The core consists of thin laminations of a magnetic metal such as iron. There is no physical connection between the primary and the secondary windings.

Choke Coils

A choke consists of a coil, generally of heavy gauge wire, wound around a fixed laminated iron core (Figure 2.28A). It is used in power supplies as part of a filter system following the rectifier.

Solenoids

A solenoid consists of a single or multilayer coil wound around a form containing a cylindrical iron core capable of moving back and forth with its movement controlled by the magnetic field surrounding the coil when a current flows through it (Figure 2.28B).

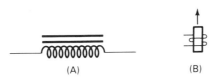

(A)   (B)

Figure 2.28. Choke (A) is an iron-core inductor. Solenoid (B) has an iron core that can move when a current flows through the coil. A spring returns the core to its initial position when the current stops.

Coil Applications

Coils are also used in the manufacture of a number of electronic components: headphones, speakers, relays, and motors. The symbols for each of these are different.

WIRES AND CABLES

If one word could be used to describe electronic components it would be "variety." Possibly nothing in electronics could be less challenging than a length of wire, and yet it is available in a large number of different styles, shapes, gauges, and applications. A wire is a conductor, and yet in some instances, such as a waveguide, it is often not recognizable as such. Sometimes special names are given to wires and are supposed to identify their specific functions, such as messenger wire, or to convey some idea of how they are designed.

Wires are known as Bell or annunciator, single-wire line, twisted pair, Litzendraht (more often called "Litz"), tinned wire, lamp cord, multiconductor cable, flat ribbon cable, mainspring cable, rollup cable, accordion cable, rotor cable, double-receptacle connector cable, extension cord, dubbing cable, single-wire transmission line, shielded and unshielded twin lead, balanced and unbalanced coaxial cable (coax), dual coaxial cable, audio cable, messenger cable, Siamese cable, microphone cable, speaker cable, Jones cable, and so on.

Despite this large number, there are just a few symbols to represent them. Cables, when used in circuit diagrams, can sometimes be identified by a symbol, or by some callout, by a combination of the two, or alphanumerically in conjunction with an associated parts lists. Often enough, there is no way of identifying the wire. In some instances it may not be necessary since the conductor(s) may be part of a printed circuit (described in Chapter 5). Sometimes a circuit may show a wire connecting two parts, and yet there may be no wire at all but simply the joining of their extension leads.

Wire Connections

A straight line is used to represent a wire for connecting components. The line may be horizontal or vertical, or, using a right angle, a combination of the two. In some cases the lines cross each other, but when they do (Figure 2.29A), no connection is intended. To emphasize this, one of the lines may form a small loop at the crossing point (B), but this technique is now rarely used in circuit diagrams. To indicate a connection, a dot is used at the point where the two lines cross.

Lines are used simply to indicate a connection between electronic parts. In some cases, even though just a single line is used, that line may represent a pair of conductors. Thus, the line between an antenna and the input to a receiver may actually consist of two wires.

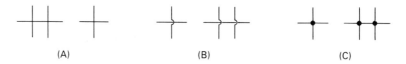

(A)   (B)   (C)

Figure 2.29.   Straight solid lines represent wires. Method of indicating no connections illustrated in (A) and (B), with (A) preferred. Dot shows connections (C).

Twin Lead

Also known as transmission line or downlead, twin lead is used to connect an FM or TV antenna (or both) to a receiver. Twin lead consists of a pair of conductors, but is often represented in a circuit diagram as a single line. No information is supplicd in the diagram as to how the two conductors are to be connected to the antenna or to the receiver.

Twin lead is also available with metallic shield braid so as to minimize the pickup of radio frequency interference (RFI). The shield is grounded either by a separate wire making contact with the braid or automatically when the cable terminates in a plug that mates with a grounded jack.

Symbols for all the wires and cables mentioned previously are either single or multiple solid lines. Shield braid is shown as a pair of dashed lines, while a solid metallic covering around a cable consists of a pair of solid lines. The shield around a single or group of conductors is always grounded, but the ground symbol isn't used. Grounding is obtained automatically when the plug terminating the cable is joined to its mating jack usually mounted on the rear of a component, such as an amplifier or other equipment.

Coaxial Cable

This transmission line and component interconnection cable is available in several general types, but are basically balanced and unbalanced. Balanced coaxial cable consists of a pair of conductors covered with a metallic shield either in braid or solid form. Each of the conductors may carry the signal, and in the case of unbalanced coaxial (coax) cable consists of a central or so-called "hot" lead and the surrounding shield (Figure 2.30A) or "cold" lead. The shield is grounded as a protection against radio frequency interference (RFI). Balanced coax consists of two wires working as signal carriers while the surrounding shield works solely as a guard against interference (B).

(A)   (B)   **Figure 2.30.** Unbalanced coaxial cable (A); balanced coaxial cable (B).

Hardline

Hardline is unbalanced coaxial cable but has a solid, outside metallic shield instead of flexible shield braid. It may use the same symbol as balanced coaxial cable but sometimes that symbol, instead of a dashed line to represent the braid, is a solid line.

Audio Cable

Commonly used to interconnect high-fidelity components it resembles unbalanced coax and uses the same symbol.

Messenger Cable

This is a combination cable consisting of unbalanced or balanced coaxial cable plus a single, independent insulated wire. The symbol consists of that used for either type of coaxial cable plus a single solid line representing the add-on wire.

Siamese Cable

Siamese cable carries the concept of messenger cable an additional step and consists of coaxial cable, either balanced or unbalanced, plus three additional wires, two of which are color-coded white, the other black. These three wires are independent; that is, each wire is insulated but all three are housed in a single jacket covering the shield braid. The black lead is a ground wire, sometimes referred to as a "drain." For a symbol (Figure 2.31), the Siamese wires are shown as three solid lines accompanying the coaxial cable symbol. Loops are used to emphasize cable construction.

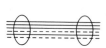

Figure 2.31. Siamese cable consists of three independent wires plus coaxial cable, either balanced or unbalanced. The loops indicate cable construction.

Multiple Coaxial Cable

This wire line consists of a pair of coaxial cables, operating independently but contained in the same jacket. Symbolically, it is shown by using a pair of coaxial cable symbols surrounded by one or more loops.

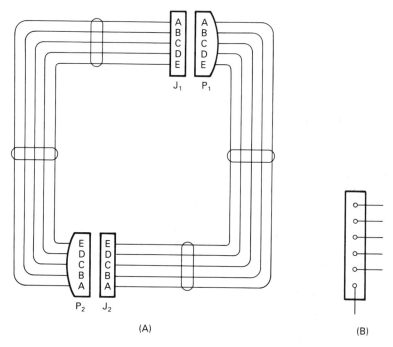

(A)

(B)

Figure 2.32. Flat cable, terminated in plugs and jacks. Plugs are identified by letter P, jacks by J. The letters indicate corresponding connections (A). Plugs and jacks for flat cable can also be represented by a symbol (B).

Flat Cable

In some instances a cable consists of wire only without shield braid or a solid shield and arranged in parallel form with that form in the shape of a flat, plastic ribbon (Figure 2.32A). Sometimes the connectors are included, and their pins are identified by letters or numbers. The letters P and J shown in this diagram are abbreviations for plug and jack. A plug has pins extending from it; a jack has holes to accommodate those pins. The four loops around the wires are cable symbols.

The drawing in Figure 2.32A is more of a pictorial than a symbol. It can be simplified by using the rectangular symbols in B. The symbol sometimes has identification letters or numbers, or both, for each of the connections.

Microphone Cables

These are members of the coaxial cable family and can be balanced or unbalanced types. They use the same symbols as those for coaxial cable but are often identified in a circuit diagram by the way in which they are used, that is, connected to a microphone or to a mixer. Sometimes a callout is used with a label such as ''mic cable.''

PLUGS

A plug, also known as a male connector, is used for connecting a wire or cable to some component, such as an amplifier equipped with a mating part called a ''jack.'' Using jacks and plugs makes it easy to join one component to another.

There are many types of plugs (Figure 2.33) including the standard 1/4-inch phone plug, phono plugs, Jones plugs, UHF plugs (also known as PL-259 plugs), F plugs, RCA plugs, BNC plugs, quick-connect plugs, XLR connectors, DIN (Deutsche Industrie Normen) plugs, 10-pin connectors, plugs for connections to printed circuit boards, and so on. For an effective connection between a pair (or more) of components, jacks and plugs must mate. They must be designed for each other. If, for example, a tuner and a preamplifier are equipped with different jacks, it is possible to use a connecting cable with different plugs on each end. Figure 2.33 E and F supplies the symbols for grounded and ungrounded AC line plugs.

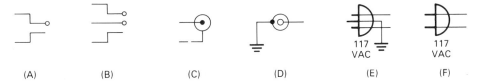

(A) (B) (C) (D) (E) (F)

Figure 2.33. Plug symbols. Two-conductor phone plug (A); three-conductor phone plug (B); phono plug (C) and phono jack (D). Grounded AC line plug (E); ungrounded line plug (F).

JACKS

While jacks are ordinarily mounted on components, they are sometimes positioned on cables as well. The symbols for various types of jacks are shown in Figure 2.34. The one in A is a phone jack and is intended for shielded unbalanced cable. That in B is for balanced cable. These are also known as open-circuit jacks.

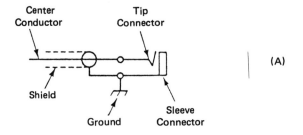

(A)

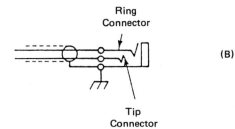

(B)

Figure 2.34. Jack for unbalanced cable (A); jack for balanced cable (B). Both of these are nonshorting (open circuit) types.

Inline vs. Open Frame Jacks

Jacks mounted on cables are known as inline jacks, and as such must be completely shielded. Jacks positioned on components such as amplifiers are open-frame types because they have no need of shielding. In some cases, however, open-frame jacks are completely shielded as a protection against unwanted signal pickup. To differentiate them from open frame types they are called "enclosed" jacks.

Shorting and Nonshorting Jacks

There are various ways of categorizing jacks: by the kind of plugs they require, by the type of cable, and whether they are shorting or nonshorting types. Figure 2.35A is the symbol of a nonshorting type. With this jack the hot lead remains floating, that is, it remains open. In the shorting jack the two upper terminals make contact, and this is "broken"; that is, it is opened, when the plug is inserted. With the shorting arrangement it is possible to put a resistive load across the hot connection, thus killing any unwanted signal pickup, such as a hum or noise voltage, by the hot lead. The unit in B is sometimes called a "closed-circuit" jack.

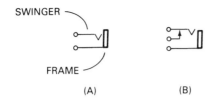

SWINGER

FRAME

(A)   (B)

Figure 2.35. Nonshorting (A) and shorting jacks (B). The jack in (A) is also called an open-circuit jack since it leaves the connected circuit open when its mating plug is removed. Inserting the plug closes the circuit.

Connectors

In some instances a connector may be nothing more than pins that make a good fit with a matching socket. These are commonly used as a plug and jack for interconnecting computer components and are the terminal connectors on flat cable. The pin contacts are referred to as male; the sockets as female. Figure 2.36A shows the symbols for both types. Figure 2.36B shows the symbol for terminals. Sometimes terminals are accompanied by some form of identification calling attention to the kind of input or output (Figure 2.36C).

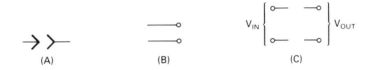

(A)   (B)   (C)

Figure 2.36. Symbols for pin-type connectors (A). Male (left); female (right). Terminals (B). These are sometimes accompanied by identifying alphabetic symbols (C).

MICROPHONES

A transducer is any device that changes energy from one form to another. A speaker is a transducer, changing electrical energy to sound energy. A battery is a transducer, changing chemical to electrical energy. A microphone, abbreviated mike or mic, is also a transducer but works in a manner opposite that of a speaker, changing sound energy to its electrical equivalent.

There are various types of microphones, and these can be identified by the way they are constructed or by the way they are used. In terms of construction, microphones are known as carbon mics, crystal, ceramic, dynamic, condenser, and electret. In terms of use they are known as omnidirectional (more often called an omni), cardioid, supercardioid, or hypercardioid.

A variety of symbols (Figure 2.37) are used to indicate microphones in general. If a specific microphone is intended, a symbol is inserted in the microphone symbol; that is, we have a symbol within a symbol. Thus a crystal mic uses the symbol for

(A)   (B)   (C)

Figure 2.37. Symbols for microphones. (A) general symbol; (B) crystal mic; (C) condenser mic. Abbreviation for a microphone is mic or mike, with mic the preferred form.

a crystal inside the microphone symbol; a dynamic mic uses a coil (to represent the moving coil of this microphone), and a condenser or an electret mic uses the symbol for a capacitor. To avoid any ambiguity, sometimes the type of microphone being used appears alongside the symbol.

HEADPHONES

Like microphones, headphones are accessory devices and are usually not shown in a pictorial or circuit diagram. However, both microphones and headphones may be included in a block in a block diagram. Components show headphone connections by a single or a pair of circles with the word "headphone" or "phone" adjacent. Neither the symbol (Figure 2.38A) nor the pictorial supply any indication as to the type of headphone.

(A)

(B)

Figure 2.38. Symbols for headphones (headset) (A) and speakers (B). The symbol at the right in (A) is used to indicate a single earpiece.

Headphones are circumaural and supra-aural. The circumaural types have cushions that cover the ear completely, thus blocking extraneous sounds. The supra-aural headphones permit hearing outside sound, but this sound is muted. With this headphone you can listen to recorded music but also hear a telephone if it should happen to ring.

Headphones come equipped with a length of connecting cord, either straight or coiled, somewhat like a spring. The cord terminates in a plug that mates with a jack on the component, possibly a receiver, preamplifier, or television set.

SPEAKERS

There are numerous types and sizes of speakers, but they can all be represented by a single symbol (Figure 2.38B), and sometimes a pictorial is used as well. Since speaker enclosures may hold two or more speakers, symbols are used for each of the speakers with a dashed line around them to indicate the presence of an enclosure. There are numerous types and sizes of speakers, as well as enclosures, but these are represented by the same symbol. The most widely used speaker is the dynamic.

SWITCHES

Switches can be operated manually, or, as in the case of relays, by an electromagnet. A relay can be regarded as a two-part device: (1) the switching mechanism consisting of the coil; and (2) the switch made up of the armature and contact points.

A manually operated switch has an armature (or blade) and one or more contacts. The simplest of these is the knife switch, also known as a single-pole, single-throw switch or SPST switch (Figure 2.39). Physically the switch has a number of different shapes, but its function is to open or close a circuit, or to permit the transfer (or stop the transfer) of a current or signal from one circuit to another and so it has a make/break action for a circuit. The on/off switch on a radio receiver or TV set is a SPST type.

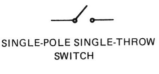

SINGLE-POLE SINGLE-THROW
SWITCH

SINGLE-POLE DOUBLE-THROW SWITCH

DOUBLE-POLE SINGLE-THROW SWITCH

DOUBLE-POLE DOUBLE-THROW SWITCH

DOUBLE-POLE DOUBLE-THROW
REVERSING SWITCH

Figure 2.39. Symbols for switches.

The SPST switch has a pair of connections, one of which leads to the blade or armature, the other to its contact point. Although the switch is a two-terminal device, it has just a single make/break contact.

The single-pole, double-throw switch (SPDT) is almost the same as the SPST type, but it has two make/break contacts. It is a three-terminal device with one connection to the blade, and the others to the two contact points. With the SPDT, the center terminal can form a connection with either outside terminal. The SPDT as well as the SPST are available in a number of arrangements, for example as slide switches, pushbutton, soft touch, and rotary.

The double-pole, single-throw switch (DPST) is like having a pair of SPST switches. In the DPST a pair of switches are operated together and can control two circuits simultaneously, turning them on or off. The unit consists of a pair of metal blades, but these are insulated from each other. In the symbol, a solid or dashed line is used to indicate the mechanical linkage between the two switches.

The DPST switch can be changed to a double-pole, double-throw by adding another pair of terminals. The two blades are insulated from each other but are mechanically linked.

Crowbar Switch

Also known as a protective crowbar, the crowbar switch is a switching circuit that functions automatically in the event an excessive load is imposed on the power supply of equipment, producing a higher than normal current flow and possibly resulting in overheating. Crowbar circuits can be made much more sensitive to deviations from normal current flow than fuses or thermal cutouts.

In some arrangements the crowbar is a discrete circuit, possibly positioned near the power supply. However, it can also be part of an associated component supplied by the power supply.

Semiconductor Switches

Semiconductor switches are characterized by the fact that they can turn current flow on or off. There are two basic types: those that control DC and those that control AC. Thyristors are switches and include silicon-controlled rectifiers for switching DC and triacs for switching AC.

Wafer Switches

In some instances it may be necessary to connect a conductor to any one of a number of contacts, possibly representing input points. The unit can be a wafer switch, and it is not uncommon for these to have as many as six contact points, or more. The rotary form of the symbol shown at the left in Figure 2.40A is more commonly used than the slide type at the right (B). The circular slide arm of the switch has a small metal extension that touches one contact point at a time as the center rotor of the switch is turned. The unit in C has 12 contact points, and so this switch could be called a single-pole, twelve-throw.

In some symbols for this switch, each contact is numbered, proceeding clockwise while facing the shaft side of the switch. In the drawing in Figure 2.40C note that the center rotary wafer is arranged so as to touch just one terminal at a time. The switch can be turned clockwise or counterclockwise so as to select any particular contact point. Wafer switches can be ganged (Figure 2.40E).

The rotor can be modified so that more than one contact point is used at the same time. Figure 2.40D shows a split ring, with each half of the ring making contact with two points.

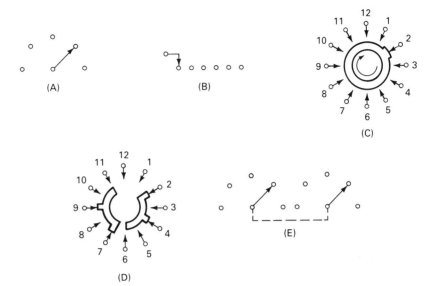

Figure 2.40. Basic symbol for the wafer switch (A) and slide switch (B). 12-contact wafer switch (C). Split stator ring wafer switch (D). Ganged wafer switch (E). The fixed position portion of the wafer is the stator; the rotating part the rotor. The dashed line indicates that the two rotor arms are ganged and are turned simultaneously.

Pushbutton Switches

Switches can be identified by the mechanical method used in operating them. A rotary switch is a type that is turned; a slide switch is one whose operating element is made to slide back and forth; and a pushbutton switch works by depressing and releasing a button. Some of the pushbutton switches in Figure 2.41 are equivalent to SPST types. Pushbutton switches can also be ganged to produce the action of DPST switches.

The refrigerator switch (Figure 2.41E) is a normally open type (NO) and is kept open as long as pressure is applied. When the pressure is removed, the switch closes, and an external circuit (possibly a light) is activated. Because of this behavior, similar to that of switches used in conjunction with refrigerator doors, it is referred to as a "refrigerator" switch. The switch is a spring-loaded type.

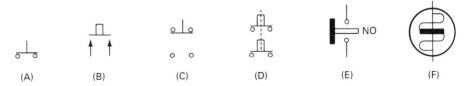

Figure 2.41. Pushbutton switches. These can be represented by either of two symbols (A or B). Both are equivalent to single-pole, single-throw types. The switch in (C) is a double-pole, double-throw. Pushbutton switches can be ganged as in (D). This is a single-pole, double-throw. Refrigerator switch (E). Mercury switch (F).

Membrane Terminal

This is a touch-sensitive type of function key and is basically a switching device consisting of a membrane-like sheet on which the key areas are imprinted.

The keypad is flat, and a light finger touch on the thin pad activates various functions. The keypad consists of three layers: two are coated with an epoxy silver contact pattern for electrical contacts. They are separated by a third layer, with holes in each of the function imprint locations.

A light touch is all that is needed for activation of this switch. The membrane layers are bonded into a specially sealed package to keep out dirt and moisture. Shielding protects the keyboard from static electricity damage.

Switch Identification

Switches can be identified as SPST, SPDT, DPST, and so on, by their construction such as knife, toggle, pushbutton, slide, or by function such as locking, magnetic, telephone, TV. Some switches are equipped with a light to indicate an on condition. Others are momentary types that keep a circuit turned on only as long as its controlling pushbutton is depressed. Toggle switches are equipped with a small plastic or metal arm extending from the switch. Mercury switches (Figure 2.41F) have a pair of metallic contacts positioned near a small pool of mercury. If the switch is moved or tilted, the mercury, an electrical conductor, covers both contacts and makes contact this way.

CIRCLE SYMBOLS

Circles are often used in circuit diagrams to represent components such as ammeters, milliammeters, voltmeters, tuning indicators, signal strength indicators, motors, relays, pilot lights, input and output points. For input and output points, the circles are quite small and are either identified by letters and/or numbers that are adjacent, or by some abbreviation.

In some instances the circles are a substitute for the actual symbol. Thus, relays have definite symbols, but in some instances the symbol may be too specific and so a circle with the letter R inscribed is used instead. In some cases, though, the circle is the only symbol available, as in the case of a pilot light (Figure 2.42).

Waveform symbols are sometimes shown at the input element of a circuit, at the output, or both. The symbol can be a drawing or an actual photo taken from the screen of an oscilloscope. In some instances, to indicate the type of input from a source such as a generator, a symbol is used showing the waveform enclosed in a circle.

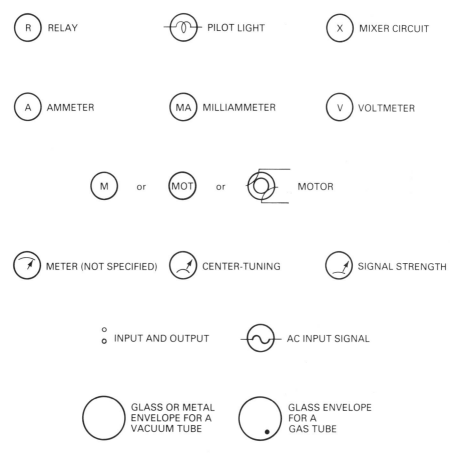

Figure 2.42. Circle symbols.

MOTORS

Motors are often considered as electric rather than electronic devices, yet they are often used in connection with electronic components. Motors are used in turntables and compact disc players. A speaker can be considered as a motor having restricted movement. There are many motor types including the induction, synchronous, DC servo, AC servo, quartz-phase lock loop (p11), outer rotor hysteresis synchronous, 120-pole linear AC servo, 120-pole linear quartz phase lock loop, brushless Hall effect, brushless, slotless DC servo, coreless DC servo 20-pole, 30-slot DC servo, and quartz DC servo. The same symbol (Figure 2.42) may be used for all these motor types, but generally the type name is placed adjacent to the symbol for identification. Manufacturers sometimes assign new names to the motors they use, but the motor is usually one of the types listed here or may have some modification to give the motor a desired characteristic.

METERS

The simplest symbol for ammeters, milliammeters, and voltmeters is a circle with a suitable letter inscribed to indicate the type of test instrument (Figure 2.42). An ammeter is a small circle with the letter A, a milliammeter uses the letters MA, and a voltmeter the letter V. In some instances just a small circle is used, and the identifying letter is placed outside. Still another symbol is a circle with an inscribed letter with a word describing the meter positioned adjacent to it.

RELAYS

A relay is an electromagnetic switch. Its basic parts consist of a coil, an iron core, and a moving arm (or armature) pivoted at one end and attached to a spring. Although a single armature is shown in the illustration (Figure 2.43), the relay can have a number of them. The relay can be a normally open type (NO), a reference to the fact that the circuit controlled by the relay is open circuited until the relay is activated. A relay can also be designed to be a normally closed (NC) type, and so the circuit controlled by the relay is operative until a current is made to flow through the coil.

There are many types of relays: a latching relay (locking relay) in which the armature is caught and held by a latching device; a stepping relay in which pulses

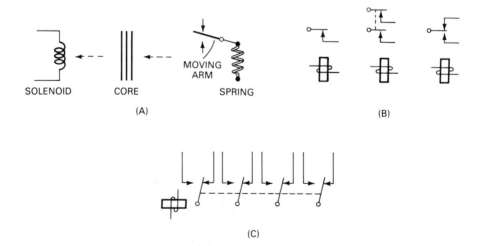

Figure 2.43. Representative relays. The coil, sometimes called a solenoid, is a fixed iron-core type (A). The relay at the left in (B) is a normally open (NO) SPST; that in the center is a NO, DPST and that at the end is a SPDT NC. The relay in (C) is a ganged unit and is a four-pole, double-throw.

of current through the relay's coil permits the armature to release a mechanical device such as a pawl or a gear; an interlocking relay equipped with two or more sets of contacts that can open or close simultaneously or sequentially, and a time-delay relay that will open or close a predetermined amount of time following the application of current to its coil.

PHONO CARTRIDGES

Also known as pickups, these components are used as transducers for tracking the grooves in a phono record, converting the mechanical movement of a stylus into an electrical signal. There are a number of different types, including the crystal, moving magnet, moving coil, and the variable reluctance (Figure 2.44).

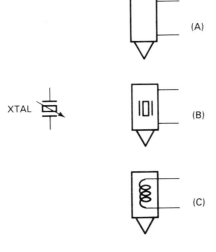

Figure 2.44. Phono pickup symbols. Generalized symbol for nonspecific pickup (A); crystal pickup (B); magnetic pickup (C). Two symbols are shown for the crystal pickup. The one at the left is more commonly used. The abbreviation xtal (crystal) may or may not be included.

FUSES

The purpose of a fuse is to protect equipment against burnout by excessive current flow. Fuses have a current rating indicated in either milliamperes or amperes. Fuses are generally found as part of the power supply. The symbols (Figure 2.45A) often

Figure 2.45. Symbols for fuses (A). The one at the top is used most often. Lightning arrester (B).

have the current rating of the fuse indicated alongside. The letter symbol for a fuse is F.

While fuses are generally associated with the AC power line input to a power supply, they are also used with speakers. The purpose of a fuse here is to protect the speaker's voice coil against burnout. This could happen if the volume control is left at some high setting, there is a strong bass tone, a frequency at which the impedance of the voice coil is very low, and a speaker shunt wiring arrangement so that the voice coil impedance combination is quite low, possibly in the order of two ohms, or even less. Figure 2.45B is the symbol for a lightning arrester.

VCR HEAD

The head in a video cassette recorder (VCR) is a coil of wire enclosed in a metal housing. The magnetic field surrounding this coil when a current flows through it is allowed to escape via a very fine gap in the face of the enclosure. It is via this gap that tape is encoded with signals or decoded. Figure 2.46A illustrates the symbol for a VCR head. The same symbol is used for all the heads, whether for recording or playback.

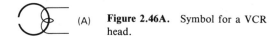

(A)   **Figure 2.46A.**   Symbol for a VCR head.

Figure 2.46B.   Symbol for a quartz crystal. The identification, xtal, may or may not accompany the symbol.

(B)

XTAL

INTEGRATED CIRCUITS

An integrated circuit consists of a number of transistors, diodes, and resistors mounted on a very small section, a "chip" of a semiconductor material such as silicon. There is no symbol for a chip, but it can be represented in a block diagram by a block. Integrated circuits are completely enclosed in an insulating material called a package. Connections to the IC are by leads coming out of the package. Chapter 5 covers this topic.

QUARTZ CRYSTALS

Quartz is a naturally occurring crystal made of silicon dioxide having piezoelectric properties capable of maintaining selected frequencies with a high order of accuracy and stability. For this reason it is used in oscillator circuits as the frequency-controlling element. In circuit diagrams the symbol (Figure 2.46B) is sometimes accompanied by an abbreviation for crystal written as xtal.

Crystals are also used in lattice bandpass filters in the intermediate frequency stages of superheterodyne receivers, and also in single-sideband receivers. The same symbol is used for the crystal for all these applications. Quartz crystals also find application in microphones, but in this case the symbol is that of a microphone with the crystal symbol inset, forming a symbol within a symbol.

Phono pickups also use a piezoelectric substance as the operating element except that it consists of a pair of thin slices of Rochelle salt. The inside symbol (Figure 2.44B) is almost the same as that used for the crystal oscillator.

OPERATIONAL AMPLIFIER

An operational amplifier, more often referred to as an op amp, is an integrated circuit consisting of a number of transistors, resistors, and possibly one or more diodes characterized by high stability and good linearity. Its symbol (Figure 2.47) consists of a triangle (A), but it is sometimes modified by including plus and minus symbols (B). A further modification appears in (C). The two inputs marked inv and noninv are references to inverting and noninverting. V_{cc} is the collector voltage; V_{ee} the emitter voltage. Each op amp has its own numbering system (D).

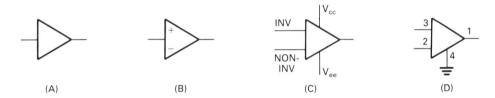

Figure 2.47. Symbols for op amps. The terminals may be numbered in any sequence selected by the manufacturer.

SEMICONDUCTORS

These devices, mostly consisting of silicon and germanium to which dopants have been added, are basically classified as diodes and transistors.

DIODES

These are two-element units having a number of applications, including functioning as mixers, rectifiers, amplifiers, oscillators, voltage dividers, demodulators, clippers or limiters, clampers and photodiodes. Depending on their function they are also listed as light-emitting diodes, Zener (avalanche), tunnel, Gunn, Schottky, Schockley, step recovery, double-base, capacitive (Varactor), backward, switching, silicon controlled rectifiers (SCRs) and so on. The general symbol for a diode (Figure 2.48A) consists of an arrow (the anode) and a short vertical line (the cathode).

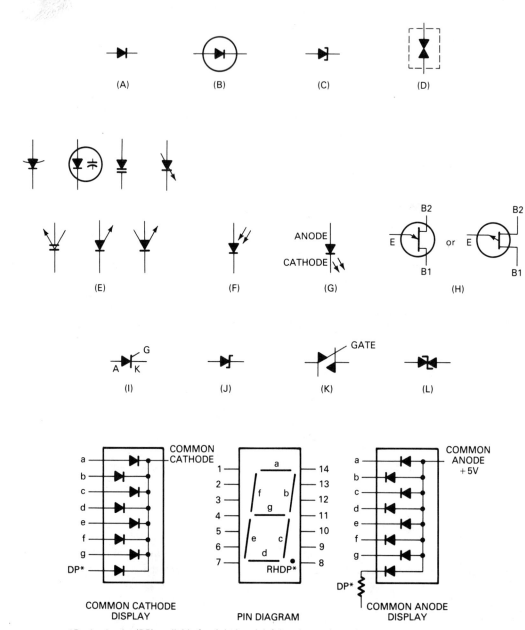

Figure 2.48. Types of diodes. Typical diode symbol used as rectifier in power supply and signal demodulator (A); tunnel diode (B); alternate symbol for tunnel diode (C); Gunn diode (D); various symbols for the capacitive (varactor) diode (E); photosensitive diode (F); photoemissive diode (G); double base diode (H); silicon controlled rectifier (I); Zener diode (J); triac (K); diac (L). LEDs (symbol G) arranged to indicate a numerical value (M).

Tunnel Diode

Unlike ordinary diodes, the tunnel diode, originally known as an Esaki diode, is not a rectifier but because of its negative resistance characteristic can work as an oscillator and also as an amplifier. Various symbols can be used, and while the circle around the diode is shown (Figure 2.48B) it is often omitted. The diode is very stable and can withstand large changes in temperature, and even when not encapsulated can tolerate moisture and atmospheric contamination. Figure 2.48C shows an alternate symbol.

Gunn Diode

This unit using gallium arsenide (GaAs) as the semiconductor material works as an oscillator at UHF and microwave frequencies. It consists of a pair of these diodes placed back to back inside a resonant cavity (Figure 2.48D), represented by the dashed lines. If a large enough voltage is applied it will oscillate, but it is not very efficient since only a small amount of the DC power applied is converted into useful radio-frequency power.

Capacitive Diode

Better known as a Varactor diode (Figure 2.48E) but also referred to as a capacitive diode, variable capacitance diode, or Varicap, the capacitative diode can be made to work as an electronically variable capacitor.

Photodiode (or Photosensitive Diode)

Unlike diodes which are encapsulated for protection, the photodiode is designed so that light can reach its P-N junction. With an increase in light intensity, its conductivity increases in the reverse direction, and so the unit can be made to work as a variable resistor. The symbol (Figure 2.48F) shows a pair of arrows pointing toward the diode, representing impinging light. The circle around the diode is usually omitted. Sometimes, in place of the arrows, the Greek letter lambda (λ) is used or else the capital letter L. A photoemissive diode (G) is one that emits light when a voltage is put across it. The symbol is the same as that for the photodiode except that the direction of the arrows is reversed. The photodiode is better known as a light-emitting diode, or LED. This electroluminescent unit emits visible, infrared, or ultraviolet light when forward biased. The semiconductor material used for LEDs is gallium arsenide, and the device is often supplied in the form of an encapsulated chip. LEDs are capable of emitting visible light in the orange, yellow, or green spectrum. These units are often used to indicate if a component or some feature of that component is functional. LEDs can be arranged to indicate some numerical value (M).

Schottky Diode

Also known as a barrier diode or a hot-electron diode, the Schottky diode is different from other semiconductor diodes in that the P-N junction consists of a metal-to-semiconductor contact, such as aluminum, and an N-type semiconductor with the metal vaporized onto the semiconductor. Its reverse bias capacitance, so useful in the Varactor, is very low. This diode can work as mixer, demodulator, or in other applications. Symbolically it uses the standard diode symbol but is identified by a callout, or alphanumerically with reference to a parts list. The design technique is also applied to transistors and are identified as Schottky transistors.

Double-Base Diode

Unlike other diodes, which are two-element devices, the double-base diode has three terminals consisting of a pair of bases and an emitter (Figure 2.48H). The unit is also known as a silicon double-base diode or silicon unijunction transistor. It has stable open circuit, negative resistance characteristics. It is made of a single N-type semiconductor material with a pair of metal contacts, the bases, connected to the ends of a silicon rod. The emitter, using P-type silicon, forms a junction on this rod, and is positioned between the bases.

Shockley Diode

The Shockley diode is a four-layer semiconductor having three P-N junctions. It is sometimes also called a four-layer or PNPN diode and is commonly used as an electronic switch. When this diode is reverse-biased, its condition is similar to that of an open switch, although there is some small amount of current flow. When biased in the forward direction, the current remains small until a certain bias voltage level is reached, and then the current increases substantially. The voltage can then be reduced while current flow remains high. The diode can be turned on or off by a pulse voltage. Switching time of the unit is just a few nanoseconds.

Silicon-Controlled Rectifier (Thyristor)

Abbreviated SCR, this is also a four-layer diode having three connection terminals consisting of a cathode (K), anode (A), and a gate (G). It is mainly used for the control of electrical power (Figure 2.48I). The unit is shown symbolically but often appears in diagrams in block form. In operation the unit is normally in an open circuit condition, but switches rapidly to conduction when an appropriate signal is supplied to the gate.

Although the word "cathode" is borrowed from vacuum-tube technology, the cathode in the SCR is not heated. The SCR is not a current amplifier (although it is somewhat similar to a bipolar transistor), but is a rectifier only, passing current in just one direction.

In operation, the anode (a word picked up from vacuum-tube technology) is made positive with respect to the cathode. This has the effect of forward biasing the P-N junction consisting of the anode and cathode. Consequently, a small current moves from the cathode, through the gate terminal, through an external load to the p-terminal anode.

The control circuit consists of the gate and the anode, and a switch in this circuit can turn the small gate-to-anode current either on or off. When this switch is closed, a much larger current flows from the cathode to the anode and then through an external load. Once the SCR is turned on, it remains turned on whether the switch in the gate/anode circuit is on or off. Operation can be stopped by opening a switch in the cathode-to-anode circuit.

SCRs can be low, medium, or high current types, with low current units capable of switching up to 1 ampere, medium current SCRs up to 10 amperes, and high current units having a switching capability of up to 2500 amperes.

Zener Diode

The Zener diode is a two-element device also known as an avalanche or breakdown diode that exhibits a sudden rise in current flow when its reverse voltage is increased above a certain point. The voltage then remains constant even with an increase in current movement. Because of this characteristic, the diode is used as a voltage regulator, voltage divider, or as a reference voltage. The symbol for a Zener diode (Figure 2.48J) is a modification of the standard diode symbol.

Triac

The triac (Figure 2.48K) is a silicon bidirectional thyristor. It consists of a pair of complementary semiconductors connected in parallel at the time of their manufacture, but using the same gate. It is found in applications such as a light dimmer for incandescent lights and in electronics as a motor speed control.

Diac

This semiconductor (Figure 2.48L) is a bidirectional current-limiting diode switch used as a variable resistance. It generally works as a trigger for a triac used as a light dimmer or motor speed control.

Clamping Diodes

Diodes can be connected as single units or in reverse parallel to form a clamping circuit. This passive network keeps the input signal, usually at some audio frequency, from going above a predetermined amount. Working as limiters, they are sometimes used in speech systems when sounds above the voice spectrum aren't wanted, with the sound power concentrated within the human voice range. When clamping results in excessive signal loss, the clamping diode network can be combined with a solid-state amplifier and in that case the arrangement becomes active, with the amplifier preceding the diode circuitry. Signals above the clamping level can show distortion. In transmitters dedicated to voice transmission, a clamping circuit is sometimes used to prevent overmodulation of the carrier wave.

Microwave Use

In microwave applications, semiconductors may be known as Impatt, Trapatt and Gunn diodes, with these designed to produce RF from DC. A PIN diode, constructed of P-type and N-type semiconductor material forming a sandwich around intrinsic (undoped silicon), can be used as an attenuator or switch. Special diodes, varactors, work as voltage variable capacitors in tuning circuits, and also in frequency multiplier units. They are designed to handle small or large amounts of current.

Diodes can also be used as temperature compensating devices, as voltage dividers, and in some cases as substitutes for resistors. They are also found in restoration circuits. As rectifiers they are extensively used in power supplies.

Diode Identification

Physically the cathode of the diode can be identified in a number of ways: by a color band at one end, the letter K, a polarity sign, or a color spot. The unmarked end of the diode is the anode. Diodes are closely related to transistors, and in some circuits a transistor can replace a diode by simply omitting any connections to the collector.

TRANSISTORS

The two basic transistor units are known as bipolar (Figure 2.49) and are the P-N-P and the N-P-N, with the letter P an abbreviation for positive, N for negative. The P-N-P and N-P-N symbols are almost identical except for the emitter connection, identified by the letter E in the drawing. This lead in is the only one equipped with an arrow and in the P-N-P transistor points inward; in the N-P-N it is outward. The other two elements in the transistor are the base, often represented by the letter B, and the collector, C.

The P-N-P and N-P-N symbols are used in circuit diagrams, but in some instances they are shown in block diagram form to indicate the direction of current flow to and through this component.

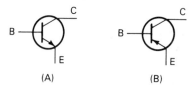

Figure 2.49. Bipolar transistors. N-P-N (A); P-N-P (B).

TRANSISTOR TYPES

As in the case of other components, transistors are supplied in a large variety of types: low-voltage, power, phototransistors, field-effect transistors (FETs), MOSFETS (metal oxide semiconductor field effect transistors), junction, unipolar FETs, coaxial, avalanche, bipolar, four-layer, hook, microalloy, planar, power, Schottky, tetrode, unijunction, gallium arsenide, epitaxial, drift field, alloy-diffused, and so on.

Special symbols may be used for some of these transistors or else the standard symbol is drawn with some identification placed near it.

Avalanche Transistor

An avalanche transistor works in the same manner as an avalanche diode, such as a Zener. It can be either an N-P-N or P-N-P type having a high value of reverse bias. A small additional amount of reverse voltage, supplied by a signal, triggers the transistor into heavy current conduction. The on/off action is comparable to that of a switch. The avalanche voltage is that amount of voltage required to trigger the transistor into conduction.

Field-Effect Transistor

Unlike bipolar transistors that are current devices, the field-effect transistor, abbreviated FET and pronounced "fett," is voltage operated. There are a number of variations of the FET including the junction (JFET) and the insulated-gate FET (IGFET) (these in turn include two variations, the P-channel and the N-channel IGFET), the metal oxide semiconductor field-effect transistor (MOSFET), the vertical metal oxide semiconductor FET (VMOSFET), and the gallium-arsenide FET (GaAs FET).

Junction FET (JFET)

At one time referred to as a unipolar transistor, this FET is a three-terminal device, and these are identified as the source(S), gate(G), and drain(D) (Figure 2.50A). The symbol for the FET shows an emitter and a two-terminal base. Current flow is from the source to the drain, and its strength is determined by the voltage on the gate. In its construction it can consist of N-type semiconductor material, sometimes called the body, surrounded by P-type, and in this case is called an N-channel FET. It is also possible to have a P-semiconductor body surrounded by N-semiconductor, producing a P-channel FET.

MOSFET

MOSFETS can be either N-channel or P-channel types, depending on which of these semiconductor materials is used as the substrate. The gate electrode is insulated from the substrate by an insulating layer of metal oxide. A conducting channel exists between this insulating material and the P- or N-type substrate. Because of this type

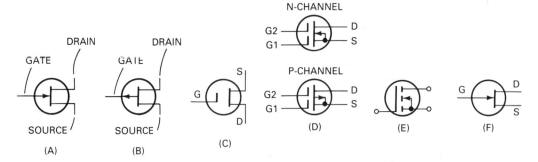

Figure 2.50. Junction FETs (A and B); the FET at (A) is an N-channel type; that at (B) is a P-channel. The direction of the arrows for the gates are opposite those used for bipolar transistors. The same symbol is used for either an N-channel or P-channel MOSFET (C). The letters G, S and D refer to the gate, source and drain. This transistor is sometimes called an IGFET (insulated gate field-effect transistor). Dual gate MOSFETS are available as N-channel or P-channel (D). Power FET (E); GaAs FET (F).

of construction, the MOSFET is sometimes referred to as an insulated gate FET (IGFET).

The input impedance of the MOSFET is extremely high, in excess of billions of ohms. As a result it imposes practically no load on a driving unit, hence does not demand input power. MOSFETS find wide application as amplifiers.

Dual Gate MOSFETS

This FET (Figure 2.50D) has two gates instead of one, as in the MOSFET described above. As such, it can be used as a mixer and a modulator, but it can also work as a high-gain amplifier.

Epitaxial Transistor

Transistors are frequently named based on the way in which they are constructed. One of these is the epitaxial, based on the fact that various layers of semiconductor material are formed or allowed to grow on a semiconductor substrate that functions as the collector. The process of semiconductor crystal growth is known as epitaxy, from which the transistor obtains its name.

The epitaxial transistor has a very thin base region, essential for high-frequency use, but the surface area at the junction of the base and collector is rather large, essential for adequate heat dissipation. Like other transistors, the epitaxial has more than one name and is sometimes referred to as an epitaxial mesa transistor or simply as a mesa.

Alloy-Diffused Transistor

The epitaxial is just one type of transistor that uses a diffusion process in its manufacture. Another is the alloy-diffused transistor, consisting of a wafer of N-type semiconductor that acts as the base with indium melted on both sides, forming the emitter and collector. With an extremely thin base region, the alloy-diffused transistor is capable of working at very high frequencies.

Electrochemical-Diffused Collector Transistor

This transistor is somewhat like the alloy-diffused, but there are differences in the manufacturing technique. However, the design here is to overcome the problem of heat dissipation in a power transistor, particularly at the junction of the base and collector.

The transistor has metal diffused into the collector using an electrochemical diffusion process. The metal is also made part of the transistor case, which, in turn, can be mounted on a heat sink.

Power FET

Also known as vertical FETs (VFETs), MOSPOWER FETs, and VMOS FETs (Figure 2.50E) this transistor can work as a linear power amplifier but also has a low signal level capability. Unlike power-type bipolar transistors, which can exhibit thermal runaway, the power FET is free of this problem. Depending on the applied bias, the VFET can be worked as a Class A, AB, B, or C power amplifier. The class of operation is determined by the amount of bias.

GaAs FET

Unlike other transistors that utilize a single element such as silicon, the GaAsFET (Figure 2.50F) uses a compound made of gallium and arsenic. It is characterized by low noise and high gain, and works well at UHF and microwave frequencies.

High Electron Mobility Transistor

Two important characteristics of transistors are electron mobility and power dissipation capabilities. High mobility is obtained by selection of a suitable semiconductor material or by eliminating dopants. Thus, gallium arsenide is better in this respect than silicon. Power dissipation can be kept to a minimum by having the semiconductor in an off-condition when it is in its non-active mode.

The high electron mobility transistor (HEMT) is also known as a modulation-doped FET or two-dimensional electron-gas FET. The HEMT is a depletion mode unit. Like other FETs it is equipped with three elements: a source, gate, and drain, built on a semi-insulating substrate of gallium arsenide. High-speed electron movement is enhanced through the use of undoped gallium arsenide. The elimination of the doping process is used to avoid hindering electron movement caused by collisions with the atoms of the dopant material.

VACUUM TUBES

For many applications, vacuum tubes have been displaced by semiconductors, but tubes are still being used in some components, either as replacements or for new equipment. In high-fidelity systems, for example, some claim that tubes produce superior sound. In some components, such as oscilloscopes and TV sets, tubes still maintain a foothold. Tubes are still being used to display voltage patterns in oscilloscopes and pictures in TV receivers, although serious efforts are being made to replace the picture tube. Tubes are still being used in electrical applications for lighting, for advertising signs, as sodium vapor lamps and fluorescent lights.

Diodes

The earliest tube ever produced was the diode, a two-element device consisting of a filament and a plate, also referred to as a heater and anode (Figure 2.51A). Diodes most commonly appear in power supplies as rectifiers and in receivers as signal demodulators.

The diode in (A) is a directly heated type. Subsequently it was supplemented by an indirect heated tube (B) using a cathode, an electron-emitting sleeve placed over the filament. The advantage of the cathode was that it separated the filament from signal-handling circuitry, a point of considerable benefit in terms of hum reduction when the filament ultimately became heated by an AC source.

Semiconductor diodes replaced tubes for a number of reasons. They were more efficient in the sense that they consumed less operating power; they required much less room, in many cases did not require sockets, and their useful operating life was greater. Subsequently, diodes became part of integrated circuits, something not possible with the tube. Tubes also required glass bulbs (metal was used later) as completely free of air as possible.

Triodes

The earliest amplifying tubes were triodes (Figure 2.51C), three-element units consisting of a filament, control grid, and plate. Subsequently, as in the case of the diode, the filament was modified to use a cathode, thus permitting the elimination of a heater battery (known as an A battery, a storage battery type). In time more elements were added to the tube. The first of these was the four-element or screen-grid tube (Figure 2.51D) and then the five-element or pentode (E) and then the beam power tube (F). Other multielement tubes followed, including the diode-triode (G) and the duplex triode (H). Other tubes included the pentagrid, electron-ray tube (the predecessor of the tuning meter), the duo diode working as a rectifier in power supplies (I), the voltage regulator tube used in power supplies to help maintain a constant output and many others. The earliest TV picture tube was an electrostatic deflection type (Figure 2.52) having a five-inch screen, followed by the electromagnetic deflection picture tube (B) still being used today.

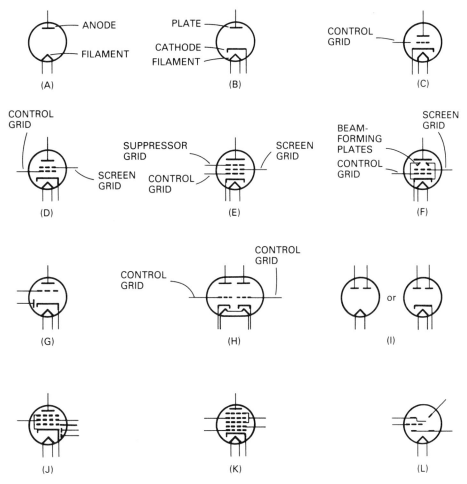

Figure 2.51. Some of the numerous vacuum-tube symbols. (A) directly heated diode; (B) indirectly heated diode; (C) triode; (D) tetrode or screen-grid tube; (E) pentode; (F) beam-power tube; (G) diode-triode; (H) duo triode; (I) directly and indirectly heated duo diodes; (J) duo diode pentode; (K) pentagrid (L) electron-ray tube. At one time this tube was used as a signal strength indicator in radio sets.

Picture Tube Numbering Systems

Television picture tubes are identified alphanumerically, with the letters and numbers not only intended to specify a particular tube type but also to supply some information about it.

The first number is the diameter of the tube in inches for tubes used in the United States or in millimeters for those intended for Japanese use. The next three letters indicate the type of tube (sometimes four letters are used), while the last letter and

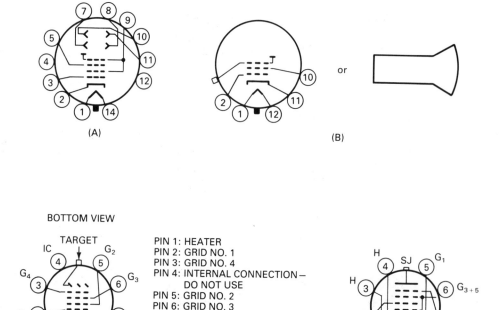

BOTTOM VIEW

PIN 1: HEATER
PIN 2: GRID NO. 1
PIN 3: GRID NO. 4
PIN 4: INTERNAL CONNECTION –
        DO NOT USE
PIN 5: GRID NO. 2
PIN 6: GRID NO. 3
PIN 7: CATHODE
PIN 8: HEATER
FLANGE: TARGET
SHORT INDEX PIN – INTERNAL
        CONNECTION –
        MAKE NO CONNECTION

DIRECTION OF LIGHT:
INTO FACE END OF TUBE

BASE BOTTOM PIN
CONNECTION FIGURE

(C)   (D)

Figure 2.52.   Picture tubes. Electrostatic deflection (A); electromagnetic deflection
(B). Symbol for a Vidicon camera tube (C). Symbol for a Newvicon camera tube (D).

following numbers indicate the phosphor. P4 is the phosphor used by a black and white picture tube; P22 the phosphor for a color tube.

A more recent numbering system is known as the Worldwide Type Designation System, abbreviated as WTDS. The first letter is either A or M, with A indicating the tube is used in a television receiver; M for a monitor. This letter is followed by two numbers and supplies the screen size in centimeters (cm). The next three letters indicate the country in which the tube will be used. Letters AAA to DZZ are for U.S. tubes; EAA to HZZ are for European tubes, while for Japan, letters ranging from JAA to MZZ are used. The two numbers following the letter code (sometimes just one number is used) supplies information about some physical difference in the tube's construction or some difference in the associated hardware. The final letter (sometimes two letters are used) indicates whether the tube is monochrome or polychrome. The letter X is used for color; WW for black-and-white.

There are also symbols for gas tubes such as the neon lamp, and the thyratron.

MISSING SYMBOLS

Symbols are available when electronic components are involved. But there are also parts that are associated with these components for which there are pictorials only. Thus, there are no symbols for a substrate material such as that used in making a transistor or a chip, or a printed circuit board. There are no symbols for electrostatic or electromagnetic fields (although a series of curved dashed lines are used for the latter); there are no symbols for temporary or permanent magnets. There are no symbols for hardware associated with electronics, such as screws, washers, nuts, and so on, but there are symbols for plugs and jacks, although these could be regarded as hardware.

There are symbols for invisible items. Thus, we can represent phantom capacitances, resistances, and inductances. Although we cannot see the flow of a current directly, it can be represented by an arrow. Voltage can be indicated by the symbol for a battery (even though no battery may be involved), or by letters such as E or V.

INDUSTRIAL SYMBOLS

Industrial symbols are used to represent parts commonly used in electronics such as antennas, capacitors, resistors, jacks, switches, relays, and so on. In some instances these symbols may be an adjunct to or part of an electronic circuit diagram, and can often lead to confusion. As an example, the industrial symbol for an air core coil resembles that of the variable resistor symbol for communications electronics. The symbol for a relay in industrial electronics looks like the electronics symbol for a capacitor (Figure 2.53).

ELECTRICAL SYMBOLS

The comments applied to industrial symbols can also be used for electrical symbols (Figure 2.54). Some of these symbols are similar to or identical with electronics symbols, as, for example, the symbol for a wire. But there are others that can be confusing. The electrical symbol for an incoming service line is the same as that used to represent an amplifier in electronics. The symbol for a telephone is that of a semiconductor diode.

ANTIQUE SYMBOLS

Some components made in the early days of radio receivers are now obsolete, and the symbols that represent them are no longer used. However, there are old-time radio buffs who look for such equipment, overhaul it, and often put it into operating order once again. There are also those who are interested in the history and technical background of electronics. For such enthusiasts the ability to read an old circuit diagram is important.

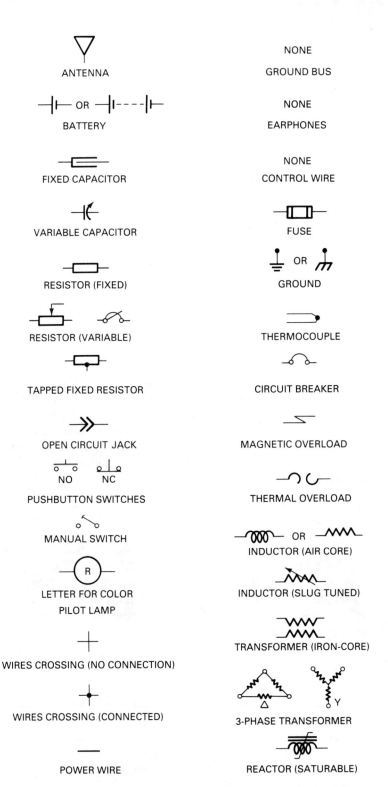

ANTENNA

BATTERY

FIXED CAPACITOR

VARIABLE CAPACITOR

RESISTOR (FIXED)

RESISTOR (VARIABLE)

TAPPED FIXED RESISTOR

OPEN CIRCUIT JACK

PUSHBUTTON SWITCHES

MANUAL SWITCH

LETTER FOR COLOR
PILOT LAMP

WIRES CROSSING (NO CONNECTION)

WIRES CROSSING (CONNECTED)

POWER WIRE

NONE
GROUND BUS

NONE
EARPHONES

NONE
CONTROL WIRE

FUSE

GROUND

THERMOCOUPLE

CIRCUIT BREAKER

MAGNETIC OVERLOAD

THERMAL OVERLOAD

INDUCTOR (AIR CORE)

INDUCTOR (SLUG TUNED)

TRANSFORMER (IRON-CORE)

3-PHASE TRANSFORMER

REACTOR (SATURABLE)

Figure 2.53. Industrial electronics symbols.

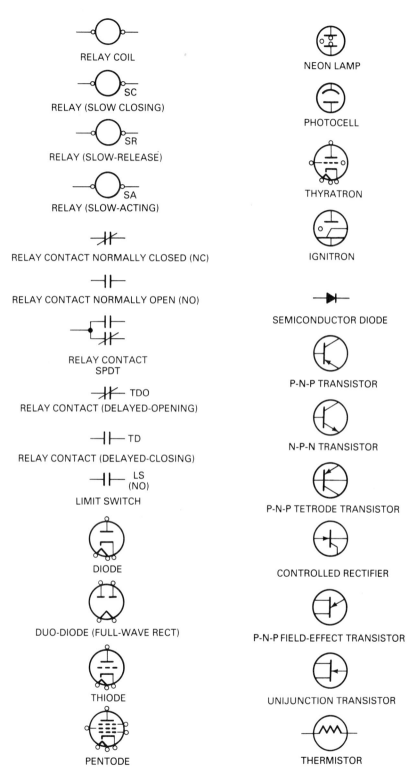

RELAY COIL

RELAY (SLOW CLOSING)

RELAY (SLOW-RELEASE)

RELAY (SLOW-ACTING)

RELAY CONTACT NORMALLY CLOSED (NC)

RELAY CONTACT NORMALLY OPEN (NO)

RELAY CONTACT
SPDT

RELAY CONTACT (DELAYED-OPENING)

RELAY CONTACT (DELAYED-CLOSING)

LIMIT SWITCH

DIODE

DUO-DIODE (FULL-WAVE RECT)

THIODE

PENTODE

NEON LAMP

PHOTOCELL

THYRATRON

IGNITRON

SEMICONDUCTOR DIODE

P-N-P TRANSISTOR

N-P-N TRANSISTOR

P-N-P TETRODE TRANSISTOR

CONTROLLED RECTIFIER

P-N-P FIELD-EFFECT TRANSISTOR

UNIJUNCTION TRANSISTOR

THERMISTOR

Figure 2.53 (cont.)

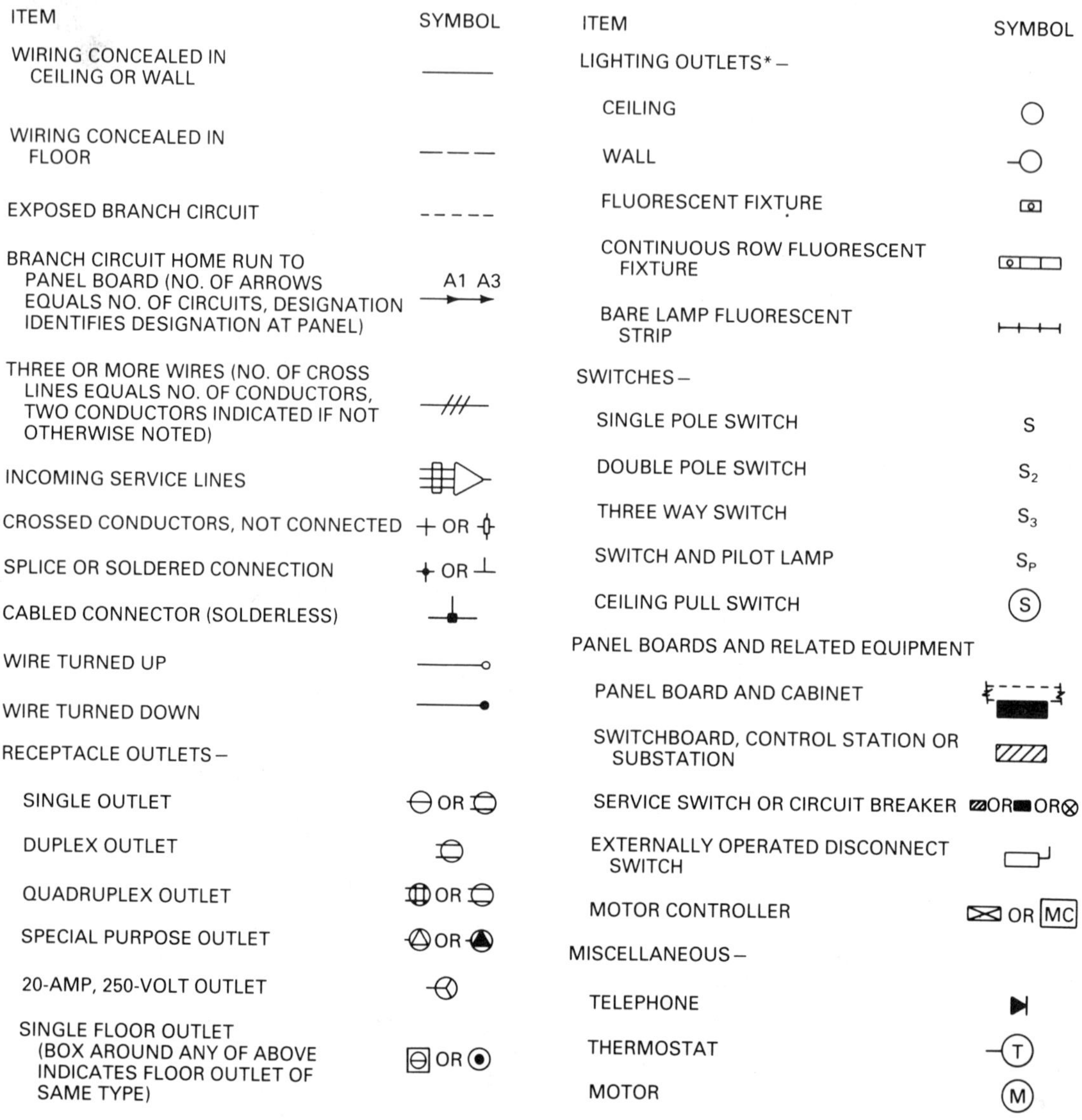

ITEM	SYMBOL	ITEM	SYMBOL
WIRING CONCEALED IN CEILING OR WALL	———	LIGHTING OUTLETS* –	
		CEILING	○
WIRING CONCEALED IN FLOOR	— — —	WALL	–○
EXPOSED BRANCH CIRCUIT	– – – –	FLUORESCENT FIXTURE	▭
BRANCH CIRCUIT HOME RUN TO PANEL BOARD (NO. OF ARROWS EQUALS NO. OF CIRCUITS, DESIGNATION IDENTIFIES DESIGNATION AT PANEL)	A1 A3	CONTINUOUS ROW FLUORESCENT FIXTURE	▭▭
		BARE LAMP FLUORESCENT STRIP	⊢⊢⊢⊣
THREE OR MORE WIRES (NO. OF CROSS LINES EQUALS NO. OF CONDUCTORS, TWO CONDUCTORS INDICATED IF NOT OTHERWISE NOTED)	–///–	SWITCHES –	
		SINGLE POLE SWITCH	S
INCOMING SERVICE LINES	⊞▷	DOUBLE POLE SWITCH	S_2
CROSSED CONDUCTORS, NOT CONNECTED	+ OR ⌀	THREE WAY SWITCH	S_3
SPLICE OR SOLDERED CONNECTION	+ OR ⊥	SWITCH AND PILOT LAMP	S_P
CABLED CONNECTOR (SOLDERLESS)	⊥	CEILING PULL SWITCH	Ⓢ
WIRE TURNED UP	——○	PANEL BOARDS AND RELATED EQUIPMENT	
WIRE TURNED DOWN	——●	PANEL BOARD AND CABINET	▤
RECEPTACLE OUTLETS –		SWITCHBOARD, CONTROL STATION OR SUBSTATION	▨
SINGLE OUTLET	⊖ OR ⊝	SERVICE SWITCH OR CIRCUIT BREAKER	▨ OR ■ OR ⊗
DUPLEX OUTLET	⊟	EXTERNALLY OPERATED DISCONNECT SWITCH	⊐
QUADRUPLEX OUTLET	⊞ OR ⊟	MOTOR CONTROLLER	⊠ OR MC
SPECIAL PURPOSE OUTLET	⬠ OR ⬟	MISCELLANEOUS –	
20-AMP, 250-VOLT OUTLET	⊘	TELEPHONE	▶
SINGLE FLOOR OUTLET (BOX AROUND ANY OF ABOVE INDICATES FLOOR OUTLET OF SAME TYPE)	▣ OR ⊙	THERMOSTAT	–Ⓣ
		MOTOR	Ⓜ

Figure 2.54. Electrical symbols.

Buzzer

The chief uses of the buzzer (Figure 2.55A) was for code practice and for testing crystal detectors to find a sensitive spot. The buzzer for either of these purposes was a special high audio-frequency type. In operation, the buzzer, a dry cell, and a telegraph key were connected in series. No headphones were necessary since the buzzer produced an adequate sound level. Code practice oscillators today are purely elec-

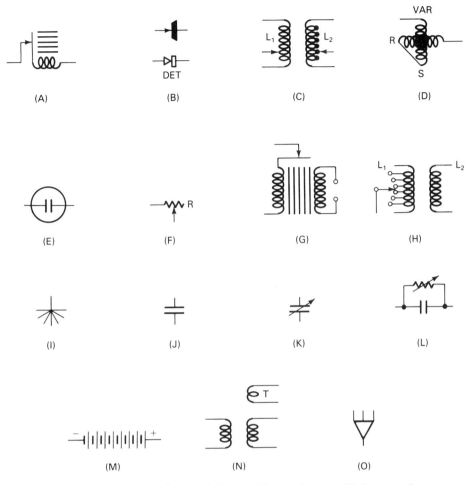

Figure 2.55. Antique radio symbols. Buzzer (A); crystal detector (B); loose coupler (C); variometer (D); chemical rectifier (E); rheostat (F); spark coil (G); variocoupler (H); counterpoise (I); fixed condenser (J); variable condenser (K); variable grid leak (L); battery (M); tickler coil (N); antenna (O).

tronic, producing a cleaner tone and using headphones. The level of the sound can be controlled, something not possible with the buzzer.

Crystal Detector

The crystal detector (Figure 2.55B) was the demodulator used in crystal radio receiving sets. There were two types: the adjustable and the fixed, with the adjustable the more popular. The crystal consisted of a small bit of galena (iron pyrites) mounted in a lead cup. Other mineral crystals included silicon, pyrites of various forms, carborundum, or a synthetic crystal.

Contact with the surface of the crystal was by a phosphor bronze wire called a "cat-whisker," whose contact point on the crystal could be adjusted.

Crystal radio sets required no outside source of electrical power and could be turned on and left on permanently. Since the crystal was a semiconductor, these early radio sets were solid state. Crystal sets supplied signal demodulation but no amplification and so relied strongly on a large outside antenna. The working range was 15 to 25 miles from a broadcast station.

Loose Coupler

One of the biggest problems of early radio receivers was poor selectivity, and not uncommonly two radio stations would be heard at the same time with the selected station stronger and the interfering station weaker, in the background.

Various schemes were tried to improve selectivity, and one of these was a variable radio-frequency transformer called a loose coupler (Figure 2.55C). It consisted of two coils, a primary coil that was fixed in position and provided with a slider that worked as a tuning device and a secondary coil arranged so that it could slide in or out of the primary, thus varying the magnetic coupling between the two coils. The secondary was also equipped with about a half dozen switch points that were connected to taps on the secondary.

Tuning, then, was accomplished by adjusting the slider on the primary winding, moving the secondary coil in or out of the primary, and by rotating a metallic arm that would move in the form of an arc over the taps leading to various points on the secondary.

The loose coupler had two distinct disadvantages. There was a certain amount of signal loss in the transfer of the signal from the primary to the secondary. And tuning required not only time, but considerable patience. However, the loose coupler, when used with a vacuum-tube receiver, was a big improvement over the crystal set.

Variometer

The variometer (or variocoupler) was the successor to the loose coupler. The advantage of the variometer (Figure 2.55D) was that it permitted faster tuning. It provided a continuously variable inductance and consisted of two coils wired in series and mounted so that one coil was able to rotate inside the other.

Chemical Rectifier

Early vacuum-tube radio receivers were operated by three battery types: (1) an A battery for supplying voltage and current to the filaments of the tubes, (2) a B battery for supplying plate (anode) voltage, and (3) a C battery for bias. The A battery was a wet type; the others were dry and were discarded after use. The A battery could be recharged. Early chargers were chemical types and consisted of one lead and one aluminum electrode immersed in a saturated borax solution (Figure 2.55E). The aluminum electrode was the positive terminal; the lead electrode negative.

Rectifier Tube

The disadvantage of the chemical rectifier was that its current output was quite small. Commonly, the rectifier was shunted across the A battery and was referred to as a trickle charger.

To accelerate charging time, a special two-element tube was used consisting of a filament and a plate. The amount of current passed by this tube was substantially greater than in the chemical rectifier.

Rheostat

Like a potentiometer, a rheostat is a variable resistor, but unlike the potentiometer consists of a wirewound element capable of carrying amperes of current. Early radio sets used a rheostat in series between the A battery and the filaments of the tubes. The purpose of the rheostat was to reduce the battery voltage to that required by the filaments. The rheostat (Figure 2.55F) was mounted on the front panel of the receiver. Some sets were also equipped with a voltmeter to indicate the filament voltage so the user could adjust the rheostat correctly.

Spark Coil

The spark coil (Figure 2.55G) was one of the early transmitters and was built so as to obtain a voltage high enough to jump across an air gap with the gap being used to send out waves at a radio frequency. The spark coil was a transformer with a primary of a small number of turns and the secondary of many turns of fine wire. This was a voltage step-up transformer with both primary and secondary wound around a laminated iron core. The current through the primary winding was interrupted by a telegraph key, with this interruption producing an alternating secondary current of very high voltage. The voltage was used to charge a capacitor (then called a condenser) until the voltage was sufficient to break down the air of the spark gap, resulting in an electrical discharge.

The spark coil produced an extremely broad wave and so transmitted a wide range of frequencies. Interference between stations was common. With the introduction of vacuum tubes, spark coil transmitters were discontinued.

Variocoupler

Because of the poor selectivity of early radio receivers, a number of components and circuits were devised to help tune in wanted stations. One of these was the variocoupler (Figure 2.55H). It consisted of a stationary primary coil (L1) and a secondary coil (L2) rotating within the primary so that the coupling between them could be changed. The primary was supplied with taps so that the coil could be tuned to the operating frequency of a selected broadcasting station. The secondary always had a fixed number of turns, and in some radio receivers further selectivity was obtained by putting a variable capacitor across this winding.

Counterpoise

In the early days of radio, the antenna and ground connections were extremely important since the sensitivity and selectivity of radio receivers was very low. In some instances, though, a good ground connection was very difficult, and in rocky soil it was practically impossible to dig deep enough to bury a flat copper plate that formed the ground. For a receiver a connection to a water pipe or steam radiator pipe was adequate, but was insufficient for transmitters.

A counterpoise (Figure 2.55I) consists of a single wire, or a group of wires fanned out, starting from a common connecting point. The counterpoise was placed far enough above ground to clear any obstructions. It was important to make sure that the counterpoise did not touch ground at any point. While the counterpoise did not make physical contact with the earth, there was adequate capacitance between the counterpoise and the earth's surface, so in effect the counterpoise was grounded for signal frequencies.

Condensers

In the early days of radio, a capacitor was called a condenser, but the name was changed since the unit didn't condense anything. Figure 2.55J and Figure 2.55K show a fixed and a variable. Initially, variable capacitors were single units, but as the number of radio-frequency stages increased, they became two- and three-ganged types. Some communications receivers used five-ganged capacitors.

Variable Grid Leak

At one time radio receivers used a battery for supplying vacuum tube bias, with the voltage supply called a C battery. This was subsequently replaced by a grid leak, a high-value fixed resistor having a range of 1 to 10 megohms. The resistor was shunted by a fixed grid capacitor. In some receivers the grid leak (Figure 2.55L) was variable, but few receivers used this arrangement. It didn't make much difference if the grid leak was fixed or variable.

Battery

The battery symbol (Figure 2.55M) consisted of alternate parallel lines of different lengths and thickness. The short thick lines designated minus; the longer, thinner lines, plus. The terminals were generally marked with plus and minus signs.

Tickler Coil

Some receivers were regenerative types, a circuit used to increase gain and selectivity (Figure 2.55N). The plate (anode) output of the detector tube was fed back to the input via a coil referred to as a tickler. The regenerative receiver was a nuisance since it could oscillate and work as a transmitter, radiating its signal and interfering with neighboring receivers.

Antenna

Because radio receivers were inadequately sensitive, long antennas were used and often consisted of wires in parallel. This is indicated by the antenna symbol (Figure 2.55O) showing three lines drawn upward from the symbol.

3

Fundamental Circuits

The development of a complete circuit diagram is a step-by-step process. It begins with symbols and then fundamental circuits as the basic building blocks. These can be assembled into subcircuits, then arranged into partial circuits, with these finally combined into complete circuits. Such circuits can be simple or extremely complex, and although they may not be instantly understandable in some cases, they can be made so by analyzing the partial or subcircuits of which they are made.

There are relatively few electronic parts, and these include resistors, capacitors, coils, operational amplifiers (op amps), integrated circuits (ICs), diodes, transistors, and tubes. A circuit diagram consists of these fundamental elements in various combinations. One of the components, not generally specified, is the wire or conductor that connects them, whether point-to-point wiring is used or a printed circuit (PC) board. There are other components, but those that have been listed here are the ones you will most often find in electronic circuit diagrams.

RESISTOR CIRCUITS

Resistors may well be the most widely used of all electronic components, and you will rarely see a circuit diagram without one or more of them. Resistors can function alone, in combination with other components, and can also be wired in three basic arrangements: series, parallel (also called shunt), and series-parallel.

Resistors in Series

One of the simplest of all circuits consists of a few resistors in series (Figure 3.1A). This diagram tells us nothing about the resistors, their ohmic value, their wattage rating, or their composition. However, some of this data is often included (Figure 3.1B).

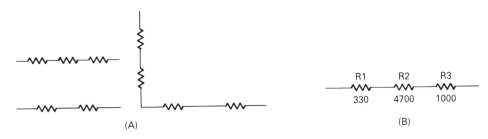

Figure 3.1.   Resistors in series.

Series resistors can be made up of fixed resistors only, or can be a combination of fixed resistors plus one or more variables. Although there is just one symbol for a fixed resistor, there are several that are used for variable types, and while there is no fully accepted standardization, the different symbols are readily recognizable.

Although resistors in series are easily identified in a fundamental diagram, they may not appear to be so in a complete circuit. In that case it may be necessary to trace the wiring in the diagram to establish the fact that the resistors are indeed in series. One possible clue is the coding. Further, in an actual component such as a radio receiver it is possible for the resistors, although wired in series, to be physically widely separated. In that case, a circuit diagram, even one with the resistors not adjacent, is helpful.

Series resistors may be drawn along a straight line, either vertical or horizontal, but in some instances may be at right angles to each other.

RESISTOR CLASSIFICATIONS

Resistors can be classified in a number of ways: by wattage rating, such as 1/4-watt, 1/2-watt; by construction, such as carbon film, carbon, metal film, metal oxide; or wirewound power types; by whether they are fixed, variable, or tapped, and sometimes by usage. They can also be arranged in the order of resistance precision: 1 percent; 5 percent; 10 percent; 20 percent, with these percentages indicating the possible plus/minus deviation from their stated values. Resistors can be characterized by the way in which their leads are connected, and for fixed resistors these can be axial or radial. An axial resistor is one that has a pair of leads, with one exiting from each end. A radial resistor has both leads coming out of one end or the other. This has no reference to quality, but the selection is determined by the ease of making physical connections.

NONSPECIFIC RESISTANCES

Every conductor, whether a straight length of wire or wound in the form of a coil, has resistance. Quite often this resistance is insignificant and can be ignored; in other conductors it can have an effect on circuit operation. With increasing frequency this resistance is sometimes an important factor.

TOTAL RESISTANCE

The total resistance of a series network of resistors is equal to the sum of the values of the individual resistors. However, this total value may not be the same in-circuit as it is out-of-circuit. In-circuit the resistors may be shunted by other components, and further, the resistors may change their stated values due to operating temperatures. If the series resistors contains one that is a variable, the total resistance will be affected by its setting.

RESISTOR POSITIONING

Theoretically, when resistors are in series it should make no difference which of the resistors comes first, second, and so on. However, in the actual construction of a component, it is often advisable to position the resistors so they are as close as possible to the other components to which they are to be connected.

TAPPED RESISTORS

A tapped resistor can be considered as a series network, but one that is fixed in position. The advantage of a tapped resistor is that no connections need be made from one resistor to another. The disadvantage is that if one section of the tapped resistor becomes defective, the entire resistor must usually be discarded, although the faulty section can sometimes be shunted with a fixed resistor. Tapped resistors are most often used where some moderate or heavy current flow is expected, as in a power supply.

SERIES POWER RATINGS

Customarily, only resistors having the same power ratings are wired in series. If, for example, four resistors are in series, with one of them rated at 1/4 watt, and the others at 1/2 watt, the 1/4-watt resistor would be the weakest link in this chain and would have a greater burnout possibility than the others. Whenever substituting one resistor for another, though, a resistor having a higher wattage rating can be used, but not one having a lower rating than the original.

COMBINED RESISTOR/SWITCHES

Potentiometers are used as volume controls (sometimes called "gain" controls) and in some instances are combined with on/off power line switches. Although these two separate components are joined on a single shaft, they usually do not appear as such on a circuit diagram. If it is essential to show that they are ganged, a dashed line will be drawn between them. If not, you will not be able to determine that they are joined components simply by reading the diagram.

PHANTOM RESISTANCE

In some instances resistance may exist in or between components even though there is no physical resistor present. To indicate that no actual resistor is used, the resistor symbol (Figure 3.2), is shown in dashed line form. Phantom resistors do not use values or codes, but are included simply to call attention to the fact that resistance does exist.

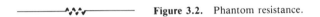

Figure 3.2.   Phantom resistance.

RESISTOR EQUIVALENT CIRCUIT

Phantom resistance can exist as an unwanted quantity between components, but it is also possible for a component to have internal resistance. In some instances the behavior of a component can be explained in terms of this resistance (Figure 3.3). Codes and values are not used for the resistors shown here. Instead the resistances may be identified as existing between the elements. Known as an equivalent circuit, it is used, in this example, to explain the operation of a transistor in a particular circuit arrangement.

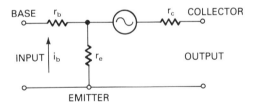

Figure 3.3.   Equivalent resistance circuit for a transistor.

INTERDEPENDENT SERIES CIRCUIT

It is possible to have resistors in series but including some other component as well (Figure 3.4). In this arrangement resistors R2, R3, and the transistor are in series. While a transistor was selected as an example in this case, other components such as coils could also be used in series with resistors.

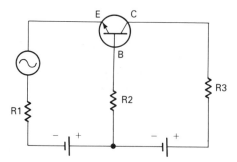

Figure 3.4. Resistors R2, R3, and the transistor are in series.

Recognizing the Resistive Series Circuit

It isn't always easy to recognize a series circuit even though it may consist of only two resistors. In Figure 3.5A, resistors R1 and R2 are in series, but this arrangement isn't immediately evident as such. However, the series network is more apparent if the circuit is pictured as in Figure 3.5B. While there is no standardization on just how circuits should be drawn, the one in A is the more usual way of depicting this circuit.

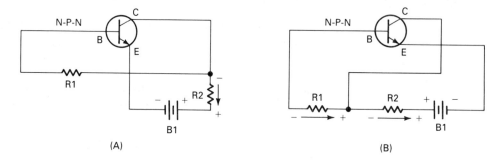

Figure 3.5. In this circuit resistors R1 and R2 are in series (A). This becomes more obvious if the circuit is redrawn (B).

RESISTIVE VOLTAGE DIVIDER

The voltage supplied by a source, such as a battery, can be divided by using a pair of resistors in series (Figure 3.6). The current flowing through the two resistors, R1 and R2, produces a voltage across each, the amount of voltage depending on the value of resistance and the amount of current flow. The sum of the voltages, called "voltage drops" or "IR drops," is always equal to the amount of voltage supplied by the source. The drawing indicates a voltage across R2, identified as e. A similar, but not necessarily equal voltage, also appears across R1.

Batteries or electronic power supplies produce a fixed amount of voltage, but with the help of resistors any value of voltage can be obtained up to the full amount

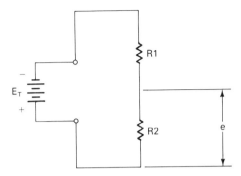

Figure 3.6. Series resistors used as a voltage divider.

furnished by the supply. A single resistor can be used to produce a required voltage drop, or the voltage divider may be equipped with a number of resistors.

Potentiometer Voltage Divider

A potentiometer is widely used as a voltage divider. Volume controls, tone controls, and loudness controls in a receiver are typical examples and are used for obtaining fairly exact values of signal voltages. In some instances the potentiometers are locking types and are factory adjusted to secure a fixed amount of operating voltage. Unlike volume controls and others of this type the shafts of these controls are located somewhere in a chassis and are not readily accessible to the user.

A potentiometer can be used when the source is either DC (Figure 3.7A) or AC such as a signal (B). In both of these illustrations a fixed resistor, R1, is put in series with the potentiometer. Its function is to limit current flow through the series network to some predetermined value.

Figure 3.7. Potentiometer as a voltage divider in a series network.

Single-Resistor Voltage Divider

Although voltage dividers often consist of two or more resistors in series, the divider can comprise just a single resistor and some other component (Figure 3.8). The single resistor in this case is identified as R1. However, this resistor is in series with the transistor and so these two components, the resistor and the transistor, form a series

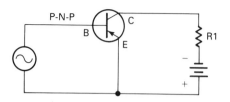

Figure 3.8. R1 and the transistor form a series voltage divider shunted across the battery.

circuit, which is shunted across the DC supply voltage, the battery. Resistive circuits can be used with a variety of voltage sources: DC, fluctuating DC, low-frequency AC such as power-line voltages, and higher frequency signal voltages.

FREQUENCY SENSITIVITY

Some parts, such as coils and capacitors, are sensitive to frequency changes, a characteristic that is not true of resistors. Except under unusual circumstances, a 100-ohm resistor has this value for DC or for AC of different frequencies. However, the value of a resistor can be affected by temperature, usually increasing in resistance with a temperature rise.

MAXIMUM AND MINIMUM RESISTANCE VALUES

The maximum value of resistance is an open circuit. Conversely, the least value of resistance is a short circuit. Either one of these conditions can afflict resistors. For the most part the net resistance of a resistive circuit is somewhere between these extremes. The word "resistance" doesn't necessarily refer to an actual resistor but can be applied to any electronic part or circuit.

CURRENT FLOW THROUGH SERIES RESISTORS

In a series resistive circuit (Figure 3.9), the same amount of current flows through each of the resistors regardless of their ohmic value. Arrows are used to indicate the direction of current flow. Since the original source voltage is DC, each of the IR drops across the resistors is also DC, with the polarity indicated. Since a voltage appears across each of the resistors, it can be used as a secondary voltage source and can then be applied to a circuit or to some electronic part. The sum of the voltages across R1, R2, and R3 is equal to the primary source voltage, a battery in this case. Although three resistors are shown here, any number can be used.

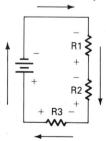

Figure 3.9. Current flow through series resistors.

RESISTORS IN PARALLEL

Two or more resistors can be wired in parallel (Figure 3.10), also known as a shunt connection. In a circuit of this kind, codes (R numbers) and values of resistance in ohms or some multiple such as kilohms or megohms can also be used. As a general rule, if the circuit shows the two resistors as adjacent, the upper resistor has the lower code number. As in series resistors, the fact that a pair of resistors are shown as adjoining in a circuit diagram does not necessarily mean they will be next to each other physically. Further, the separation may be such that the fact that the resistors are in parallel may not be immediately evident.

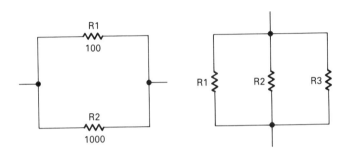

Figure 3.10. Resistors in parallel.

In a circuit, resistors in parallel can be arranged so that, at first glance, they look as though they are in series (Figure 3.11). Here it is necessary to be careful since the connections to the shunt resistors may be correct or may result in a short circuit.

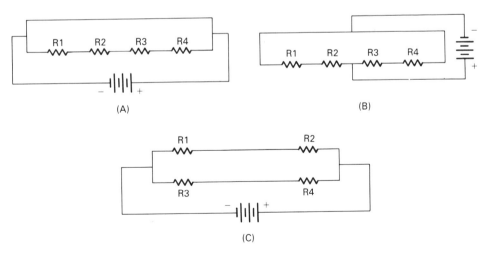

Figure 3.11. Resistors are short-circuited (A). With a different connection resistors are in series-parallel (B); equivalent circuit makes resistor arrangement more easily recognizable (C).

Whether resistors are wired in series or in parallel can depend on how the lead-in connections are made (Figure 3.12). This drawing shows three resistors wired in a triangular configuration. If, as in A, the lead-ins are connected to points A and B, then resistors R2 and R3 are in series, and this series network is shunted by R1. If (Figure 3.12B) wire connections are made to B and C, then R1 and R3 are in series, shunted by R2. Finally, as in Figure 3.12C, with connections made to A and C, R1 and R2 are in series and are shunted by R3.

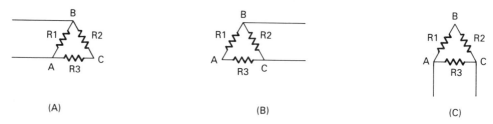

Figure 3.12.   R1 is in parallel with R2 and R3 in series (A); R2 is in parallel with R1 and R3 in series (B); R3 is in parallel with R1 and R2 in series.

WATTAGE RATING FOR SHUNT RESISTORS

When resistors are wired in series, each usually has the same wattage rating. However, those used in a parallel circuit can have independent ratings. And, like series resistors, any number of them can be used. The resistors can all be fixed types, or can be some combination of a fixed and a variable unit.

TOTAL SHUNT RESISTANCE

When resistors are wired in series, the overall resistance is increased. The opposite is true when they are wired in parallel. With shunt resistors, the total resistance must always be less than the lowest value of any of the individual resistors. Thus, when two resistors of equal value are wired in parallel, the equivalent resistance is equal to one-half that of either one. As an example, two 100-ohm resistors in parallel are equivalent to a single resistor of 50 ohms. Technically, circuit conditions permitting, the pair of 100-ohm resistors could be replaced by a single 50-ohm resistor.

Resistors in parallel need not have identical resistances and can have any ohmic value wanted. The overall effective or equivalent resistance can be calculated by taking the product of the resistance values and dividing this by their sum. In terms of a formula we have total resistance = (R1 × R2)/(R1 + R2). As an example, the total resistance of 40-ohm and 30-ohm resistors in parallel equals (40 × 30)/(40 + 30) = 1200/70 = 17.14 ohms. Note that this resistance is lower than the value of either resistor.

If three resistors are in parallel, the equivalent resistance can be obtained by using a two-step process. Take any two of the resistors and combine them into one using the method described above. Then combine the equivalent resistance with the value of the remaining resistor.

Current Flow Through Shunt Resistors

Current at the junction of two resistors in parallel divides inversely in proportion to the value of the individual resistors. The greater the resistance, the smaller the current flow. The sum of the currents through the two resistors is equal to the amount of current flow leaving the DC voltage source. The arrows (Figure 3.13) show the current paths. The same amount of voltage appears across each of the parallel resistors and is equivalent to the amount of battery voltage.

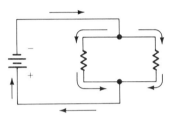

Figure 3.13. Current flow through shunt resistors.

The same analysis applies to three or more resistors in parallel. The voltage across each resistor will be identical to the source voltage even though a different amount of current will pass through each resistor, assuming they have different resistance values. The amount of current leaving the battery will be exactly equal to that returning to the battery.

Resistor Switching

Individual resistors, or resistors wired in series or parallel, can be switched in or out of a circuit. While a potentiometer could be used, individual resistors may be wanted so as to always supply a known, fixed amount of resistance.

RESISTORS IN SERIES-PARALLEL

The two basic resistor arrangements, series and parallel, can be combined to form a series-parallel circuit (Figure 3.14). The units can be fixed, tapped, or variable, and can be any combination of these. Circuits can use resistor symbols without further identification, or they may have both codes and values. While connections can be made at the ends of the series-parallel network, either taps or connections can also be made anywhere along the line.

The resistor symbol can be used not only for actual resistors but for any component that exhibits resistor behavior. In an equivalent circuit, the resistor symbol can be used to indicate a transistor. A coil of wire contains resistance and so may be represented by both the inductor and resistor symbols. A resistor is sometimes

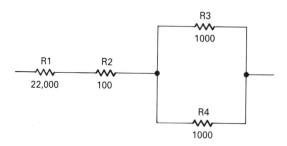

Figure 3.14. Resistors in series-parallel.

used to represent a load on a voltage supply, and that load is any device requiring current from that supply.

Because resistors in series, parallel, or series-parallel may be widely separated physically, the actual circuit configuration may either not be recognized or misinterpreted. In that case, as indicated previously, it may be helpful to trace the current path or to redraw the circuit in a more simplified form.

Since resistors are so widely used, a component such as a receiver or an amplifier may have a number of series, parallel, or series-parallel circuits, with these including parts other than resistors. In all instances, the current flow must have a clear, unbroken path to and from the voltage source.

CIRCUITS IN SERIES-PARALLEL

It is very easy to recognize a series or a parallel or a series-parallel circuit if these consist only of resistors and if the symbols are placed adjacent to each other. It is quite another matter to look at a circuit diagram and to recognize it as consisting of similar groups.

To do so, select a convenient starting point such as the negative terminal of a battery, or the positive or negative voltage input points if these are shown, or the output terminals of an electronic supply. The selected path must be one that is complete, starting at the voltage source and returning to it.

To do a complete analysis, each part in the diagram must be included, and so it is generally necessary to have a number of paths. As a result, some parts may be included more than once.

The circuit in Figure 3.15 is that of a single transistor amplifier. To see what this circuit is really like, start at the minus terminal of battery B1 and move to the left to the bottom of R1. Our objective is to find a path back to the plus terminal of the battery. We can move up through R1 to R4, through this resistor to R3, and then through this resistor to the battery's plus terminal. This supplies us with several important bits of information. The three resistors, R1, R4, and R3 are in series, and this series combination is shunted across the battery. Further, because this is a series network, the same amount of current will flow through each resistor, and finally, each resistor will have a voltage drop across it with the polarity indicated. The amount of voltage will depend on the ohmic value of each resistor and the sum of these voltages will be equal to that supplied by the battery.

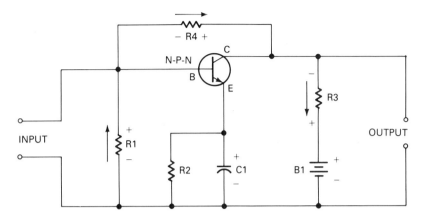

Figure 3.15. Transistor amplifier has components in series and in parallel.

This first path does not include all the resistors or the transistor, and so a second path is required. To include these parts move from the minus terminal of the battery to the bottom of R2, through the transistor from the emitter (E) to the collector (C) and then through R3, arriving at the plus terminal of the battery. This means that R2, the transistor, and R3 are also in series. We can regard the transistor in the same light as a resistor since there is a voltage across it and current flows through it.

The only part not covered in this analysis is capacitor C1. It is shunted across R2. The current flowing through R2 produces a voltage across it, and this voltage charges C1. Its function is to maintain a steady voltage across R2 in the event it should fluctuate with the arrival of a signal at the input.

These relationships are shown more clearly in Figure 3.16. This arrangement emphasizes that the original circuit is just a series-parallel combination. R1 and R4 are in series, and so are R2 and the transistor. These two series combinations are shunted across each other, and they are in series with R3. Finally, the complete series-parallel circuit is shunted across the battery.

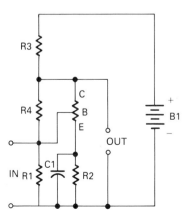

Figure 3.16. Parts relationships of Figure 3.15.

There are still two more shunt circuits to consider. The first is that the input signal is in parallel with R1. This is interesting since R1 is part of the series resistive network consisting of R1, R4, and R3. So a radio part, as in this example, can be involved in a parallel and a series arrangement at the same time.

One terminal of the output is connected to the collector of the transistor, with its other terminal to the bottom end of R2. But R2 and the transistor are in series while the output is shunted across these two parts.

The diagram in Figure 3.16 is different because it does not resemble that of Figure 3.15, the much more commonly accepted way of drawing this circuit. But Figure 3.16 is helpful in analyzing the circuit, showing as it does that it consists of series-parallel connections of a few radio parts.

METER AMPLIFIER

While transistors are commonly used as signal voltage amplifiers, they have numerous other applications. One of these is as a meter amplifier (Figure 3.17), and in this case the transistor gives the meter an apparent ten-time increase in sensitivity. Although the meter used here has a full-scale deflection of 1 milliampere, (a thousandth of an ampere), with the help of the transistor it becomes capable of reading currents in the order of microamperes (millionths of an ampere).

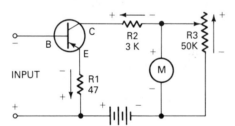

Figure 3.17. Meter amplifier.

To analyze the circuit it has been redrawn as in Figure 3.18. Start at the minus terminal of the battery and proceed through R3, R2, the transistor, and then R1 to arrive at the battery's plus terminal. The same amount of current flows through all these parts, producing voltage drops.

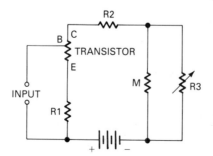

Figure 3.18. Equivalent circuit of meter amplifier.

Since any circuit analysis must cover every radio part, another trip is required so as to include the meter. Start at the battery's minus terminal, go through the meter (M) through R2 and the transistor, through R1 and then to the plus terminal of the battery. The meter has a voltage across it and current through it and so it can be represented as a resistor. This second trip shows that it is also a series network.

BATTERIES

A battery is a DC voltage source used for supplying electrical energy to circuits. The battery symbol is used not only to indicate the use of one or more batteries, but may represent any DC voltage source, including electronic power supplies operated from the AC power line.

The current flow from a battery, unlike that from an AC power source, is always in one direction. It is a convenient fiction to assume that the movement of current starts at the negative terminal of a battery, moving through a connected circuit and then back to the positive terminal. Inside the battery, current flows from the positive to the negative terminal.

In flowing through the parts that constitute a circuit, the current from the battery can follow one or more paths. In doing so the current may divide, not just once but a number of times, but in all instances, no matter how many times this division takes place, the same amount of current leaving the battery returns to it. No current is ever lost.

The direction of current flow is indicated by one or more arrows. Ordinarily these are omitted, but are used if required for some special purpose. The tail end of the arrow is always minus; the symbol indicating the head of the arrow is always plus. These are the same polarity symbols that are used when drawing a battery symbol.

BATTERY CONNECTIONS

In some components, even those having several circuits, just a single battery may be used, but others have arrangements requiring two or more. Like resistors, batteries can be wired in a number of different ways: series aiding, series opposing, parallel, and series-parallel.

Batteries in Series Aiding

When batteries are wired in series aiding, the total available voltage is equal to the sum of the voltages of the individual batteries (Figure 3.19). Generally, there are two requirements for a series-aiding connection. The first is that identical batteries should be used. The second is that both should have the same current-delivering capabilities. The fact that a pair of batteries are the same does not mean both can supply equal amounts of current. One of the batteries may be fresh, fully charged; the other practically discharged.

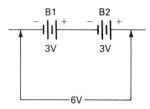

Figure 3.19. Batteries in series aiding.

Batteries in Series Opposing

In some circuits a pair of DC voltages may be arranged so they oppose each other. If the batteries have equal potentials, the equivalent voltage across their output will be zero. If the voltages aren't equal subtract one from the other (Figure 3.20A). A setup of this kind is used in control circuits in which one DC voltage is balanced against another. Usually batteries are not intended for this purpose. Resistive voltage drops are used instead. (Figure 3.20B). Not only is this more practical, but the currents through the two resistors can be controlled more precisely.

(A)   (B)

Figure 3.20. Batteries in series opposing (A). Voltage drops across resistors can also be in series opposing (B).

Batteries in Parallel

Batteries supply both voltage and current. To increase the amount of voltage available, batteries, as indicated above, can be wired in series aiding. The amount of current that can be delivered by a battery, known as its ampere-hour rating, is the amount of current that can be supplied over a given period of time.

If the amount of current that can be supplied by a single battery is inadequate, another battery can be connected in parallel with it (Figure 3.21). The same precau-

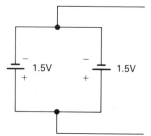

Figure 3.21. Batteries in parallel. Single batteries are sometimes referred to as cells.

tions used for batteries in series should be followed for the shunt connection. The batteries should be identical. When a moderately weak battery is wired in parallel with a more fully charged unit, the stronger battery will charge the weaker, but the current required for this charging operation will take precedence over the current to be delivered to the load.

Batteries in Series-Parallel

The fact that resistors may be wired in series-parallel does not establish a precedent for batteries. For batteries, a series-parallel connection (Figure 3.22) is a matter for consideration only if both a higher current capability and a higher terminal voltage are required. The batteries must be identical types.

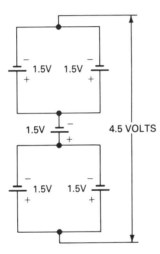

Figure 3.22. Batteries in series-parallel. Each of the batteries is rated at 1.5 volts.

Internal Resistance of a Battery

Every battery has a certain amount of internal or phantom resistance. This resistance is not fixed, but is variable and is dependent on the age and state of charge of the battery. When the battery is fully charged, the internal resistance, never completely eliminated, is at its minimum. As the battery is used, and as it gets older, its internal resistance increases. Finally, in time, charging can no longer lower internal resistance and the battery is no longer useful.

The current that flows out of the battery must pass through its internal resistance, but this internal resistance is in series with any load external to the battery. And as battery resistance increases, less and less current becomes available.

If batteries are connected in series, the limit of current availability is determined by that battery having the highest internal resistance.

LIVE AND ACTIVE CIRCUITS

Any circuit or any complete component that is actively receiving electrical power is live, whether that power is supplied by a battery or comes from an electronic power supply connected to the AC line. A receiver or amplifier or any other component that has its power switch in the on position and is connected to an electrical voltage source is live, and so are all the voltage operated circuits in that component.

An active circuit is one that is connected to a transistor or vacuum tube. It is referred to as "active" as a descriptive term and is considered an active circuit whether or not it is in the process of receiving power. Technically, a transistor or tube, or an amplifier using these parts, are always active types. They are both active and live when receiving electrical energy.

BATTERY RESISTOR CIRCUITS

With just these two components, resistors and batteries, it is possible to develop a number of live circuits, those which have current flow. These range from a single resistor shunted across a battery to a series-parallel resistor group connected across a combination of batteries.

SWITCHED CIRCUITS

Circuits can be opened or closed by switches, quite often a single-pole, single-throw type (Figure 3.23). With this circuit, current flows only when the switch is closed. However, a voltage can be measured across the terminals of the open switch and the amount measured will be very close to that of the voltage source. With the switch closed, no voltage will appear across it and the load, represented by the resistor, will have the full amount of supply voltage across it.

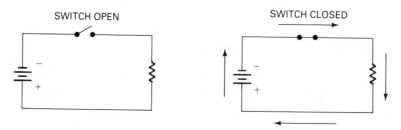

Figure 3.23.   Circuit switching.

A large variety of switches, mechanical, electromechanical, and electronic is available for switching any number of circuits on and off, simultaneously or sequentially. Possibly one of the most common types is the mechanical SPST (single-pole, single-throw) used in conjunction with a volume control for radio receivers. These two parts are operated by the same shaft, but each has its own symbol most often widely separated on a circuit diagram. No dashed line is ever used, so the diagram does not reveal the mechanical linkage.

The single-pole, double-throw (SPDT) switch can be used to permit the selection of two different signal sources as the input to a transistor (Figure 3.24). Switches of this kind are also used for in-home TV systems to provide a choice of two different video signals, possibly broadcast TV and a video game. They are also available in a variety of forms: toggle, slide, pushbutton, and rotary.

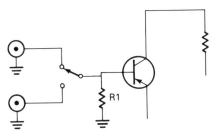

Figure 3.24. SPDT switch permits selection of two different input signals, with the signal placed in parallel with R1.

Wafer switches (Figure 3.25) are used to connect radio parts so they form a variety of circuits. The illustration is that of a three-pole five-throw type. When the rotary arm is in its first position, resistors R1, R4, and R5 are connected in parallel. The connection is broken when the arm is moved to any of the other positions. This unit is a triple-wafer switch, but some have two or more wafers. For a multiple wafer switch, the connections and circuit switching can become quite complicated.

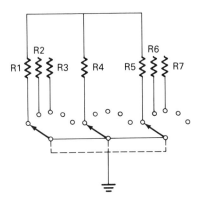

Figure 3.25. Wafer switches can be used to form a variety of series, parallel, or series-parallel circuits.

RELAYS

The switches in Figures 3.24 and 3.25 are operated manually, but can be combined with a relay with the on-off action controlled by a current (Figure 3.26). The circuit is that of a light-operated relay. When light shines on the solar cell, a current will flow through the transistor. This current will move through the relay making its coil into an electromagnet. The electromagnet will attract the armature or moving arm of the relay, and so it will be pulled down from the position shown in the drawing.

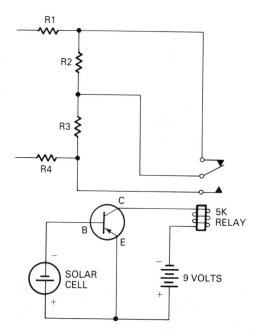

Figure 3.26. Relay works as a SPDT switch. When in the up position R2 is removed from the circuit. When in the down position R3 is removed. The circuit is not opened but the total series resistance is changed.

When the light striking the solar cell is removed, current flow through the transistor and through the relay coil stops. The armature of the relay is then returned to its original position by a spring. This action is that of an SPDT switch.

Like wafer switches, relays can be quite complex and can have a number of moving arms working as two or more switches simultaneously. Relays can be connected in parallel (Figure 3.27) or in series. In some circuits, one relay is used to operate still another relay (Figure 3.28).

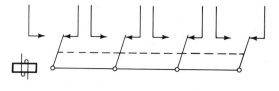

Figure 3.27. Parallel connection of relays.

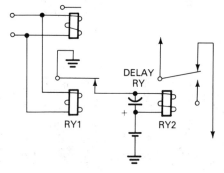

Figure 3.28. When RY1 closes, current is delivered to RY2 and it closes. When RY1 opens, current to RY2 is stopped but RY2 remains closed until its shunt capacitor discharges.

TRANSISTOR SWITCHING

Transistors and diodes can be used as electronic switches. If, for example, a transistor is biased so that no current flows through it, we would then have the equivalent of an open switch. With the transistor cut off, no current would flow through its load. We could then input a pulse to the transistor to take it out of cutoff, and this condition would then continue for the time duration of the pulse. We would then have the equivalent of a closed switch. With the ending of the pulse, the transistor would revert to its open-circuit condition.

We could also have the reverse situation. The transistor circuit could be such that current would flow through it, representing a closed-switch condition. An input pulse could drive the transistor to cutoff, supplying the equivalent of an open switch.

Figure 3.29 shows one circuit possibility. The operating condition of the transistor is set by variable resistor R2, shunted across the battery. The setting of this resistor can be such that current flow through the transistor is maximum, comparable to having a closed switch. The load in this example is an inductor, L1, but it could be any other load.

The input could be a succession of pulses having an amplitude great enough to overcome the bias supplied by R2. These pulses could put the transistor into a current cutoff condition, with no current flowing through L1. The on/off time of the transistor could be controlled by the time duration of the pulses.

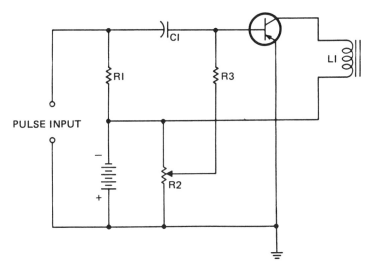

Figure 3.29. Transistor can be switched on or off by input pulses.

DIODE SWITCHING

Like a transistor, a diode can be biased so that it remains cut off until such time that an input pulse takes it out of cutoff. With suitable input pulsing, the diode can be made to work like a mechanical SPST switch.

Each type of switching, mechanical, electromechanical, or electronic, has its advantages and its shortcomings. Mechanical switching is used for external control of a component. On an amplifier, for example, a knob-controlled mechanical switch is used to make a selection of an input source such as a turntable, compact disc player, tuner, and so on. Electromechanical or electronic switches are used inside a component with operation depending on circuit conditions.

REFERENCE POINTS

All voltages are developed across two points, either one of which could be used as a reference. In the series circuit of Figure 3.30 there are two resistors, R1 and R2, connected in series, and with a current flowing through them there is a voltage drop across each. Point A on R1 is minus with respect to ground, positioned at the junction of the resistors, assuming that ground has been selected as the reference point. Point A is also minus with respect to point B if that point is the reference. In a similar manner, point C is positive with respect to ground or point A.

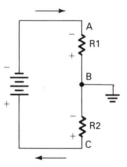

Figure 3.30. Ground used as the reference point.

The ground connection itself is either positive or negative depending on whether point A or point B has been chosen as the reference. Ground is positive with respect to point A and is negative with respect to point C. Thus, ground is either positive or negative depending on the selection of the reference.

Ground is often selected as a reference point, not that it has any special value or significance, but simply because it is convenient to do so. Ground can be an actual connection to the earth, either directly or indirectly, or it can be a connection to a "floating" ground, possibly a length of wire used for a common connection by various components. This wire is not joined to earth in any way, hence it is categorized as "floating."

Biasing by Reference

The amount of conductivity, ranging from zero to maximum, is determined by the biasing voltage placed between the base and emitter of a transistor. In Figure 3.31 the emitter in both drawings is grounded, and so is a selected voltage point on the

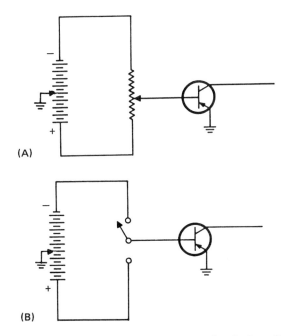

(A)

(B)

Figure 3.31. Transistor on-off operation can be determined by a potentiometer (A) or by a SPDT switch (B).

battery. In drawing A the amount of polarity of this bias can be adjusted by the setting of the potentiometer control. Thus the base can be made positive or negative with respect to the emitter.

Instead of using a potentiometer, Figure 3.31B uses a single-pole, double-throw switch. In one position the transistor is biased for maximum current flow; in the other position for current cutoff.

CAPACITORS

Like resistors or batteries, capacitors can be wired in series, in parallel, or in series-parallel combinations (Figure 3.32). When capacitors are in parallel, the total capacitance is the sum of the individual units. In series the formula is the same as for resistors in parallel, that is, C = (C1 × C2)/(C1 + C2).

Capacitors are available in a wide choice of types. They can be categorized as fixed or variable; by construction (metallized, paper, mica, tantalum, ceramic); by shape, such as disc; or by usage, such as filter, coupling, or bypass. Variable capacitors can be referred to as tuning, trimmer, single, or ganged. Capacitors can be used alone,

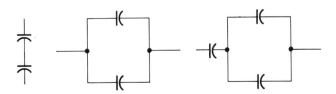

Figure 3.32. Capacitors in series (left); in parallel (center) and series-parallel (right). The parallel arrangement is sometimes called shunt.

or in some combination with resistors, coils and DC voltage supplies. Next to resistors, capacitors are possibly the most widely used of all components. Capacitors, such as electrolytics, are polarized, and this is indicated by plus signs, or plus and minus signs, on the capacitor symbol.

Capacitors can also be categorized by the amount of their capacitance, expressed in microfarads or picofarads, possibly including their DC working voltage as well. In a circuit diagram the working voltage is usually omitted, but the amount of capacitance may be included, expressed as μF for microfarads; pF for picofarads. Whenever possible, capacitance is expressed as a whole number in preference to a fractional decimal.

VOLTAGE VARIABLE CAPACITORS

Voltage variable capacitors are known under a number of different names such as varactor, parametric amplifier diode (or paramp), semicap, and varicap. Of these, varactor is the most widely known.

Voltage variable capacitors belong to the general family of parametric devices and take advantage of the voltage variable capacitance of a voltage-variable P-N junction. The symbol for a varactor is similar to that of a solid-state diode, and it is inscribed in a circle with the letter C to show it is functioning as a capacitor. Sometimes the capacitance value is included. The varicap can be used in the front end of TV tuners, or for radio receivers, or in an oscillator for modulating an audio signal for an FM transmitter. In the fundamental circuit of Figure 3.33 a capacitor C1 is in

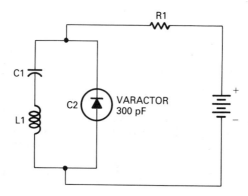

Figure 3.33. Varactor tuning. C1 and L1 are in series with this combination shunted by C2. R1 is in series with the DC power source and is, in turn, shunted across the C1, L1, C2 network.

series with a coil, L1. These two components are shunted by a varactor diode. C1 and L1 form a fixed tuned circuit. Actual tuning is done by a varactor, C2, with its capacitance varied by the DC voltage applied through R1.

There are many other functions for capacitors, including filtering, coupling, bypass, DC blocking, and so on.

COUPLING

A capacitor can be used to help in the transfer of a signal from the output of one stage to the input of next (Figure 3.34). The capacitor blocks DC voltage from being applied to the preceding stage and so it performs a double function: coupling and blocking. C1 and C2 in this figure do both.

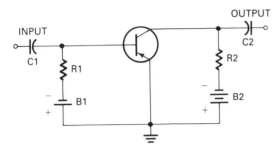

Figure 3.34. C1 and C2 are coupling capacitors but also block the DC voltage supplied by B1 and B2 from reaching the preceding and following stages.

BYPASS

In the fundamental circuit (Figure 3.35) two currents are simultaneously present at point A: DC and AC. The DC flows through the resistor but does not pass through the capacitor. Most of the AC is shunted around the resistor. This action could be referred to as passing and blocking.

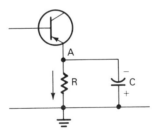

Figure 3.35. Capacitor C shunts resistor R. The presence of C results in a smoother, steadier DC voltage across the resistor.

TUNING

A capacitor shunted by or in series with a coil is a resonant circuit and can be called an adjustable tuned circuit when either the capacitor or coil (or both) are variable. The tuning range is determined by the maximum and minimum capacitance of the variable, and by the inductance of the coil. The tuning element can be a variable capacitor, a varactor diode, or a variable inductor.

PHANTOM CAPACITANCE

Capacitance can exist even in the absence of a physical capacitor (Figure 3.36). All it requires is a pair of separated conductors. The material between the conductors, known as the "dielectric," can be any substance, gas, liquid, or solid. Air is a dielec-

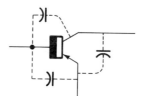

Figure 3.36. Representation of the phantom capacitances of a transistor.

tric, and so is paper, mica, or plastic. A pair of adjacent wires in a receiver, tuner, or amplifier works like a capacitor, and, as in the case of resistors, is a phantom component. As such, its symbol will be shown in dashed form. Unless there is some special reason for doing so, neither phantom resistance nor phantom capacitance is shown in a circuit diagram.

Phantom resistance is not affected by frequency and remains the same. Phantom capacitance, though, is frequency responsive, and the higher the frequency, the more effective it becomes. For high-frequency circuits, such as those used in broadcast or satellite TV, phantom capacitance can be a serious problem. In a high-fidelity system it can reduce high-frequency response and in television it can weaken picture detail.

COILS

Coils, third in the family of circuit components, like resistors and capacitors, are available in a wide variety of sizes, shapes, and functions. In terms of structure, a coil can be an air-core type, have a fixed iron core, or a movable polyiron core. It can be designed for low-frequency or high-frequency use. It can be part of a tuning circuit or a circuit having a fixed resonance. It can be part of a transformer, or an antenna system, or a relay, or any one of a number of other applications.

Like resistors and capacitors, coils can be connected in series, but unlike these other components, coils are surrounded by a magnetic field when a direct or alternating current flows through them.

The inductance of a coil (L) is measured in henrys. Submultiples are the millihenry (mH) or thousandth of a henry, and the microhenry (μH), or millionth of a henry. When two coils are wired in series, the overall inductance is the sum of the inductances of the two (or more) coils, provided their magnetic fields do not interact (Figure 3.37A). This can be done by shielding the coils or separating them sufficiently. The formula for total inductance in this case is L = L1 + L2.

If the coils are arranged so they are fairly adjacent and so that current flows through each coil in the same direction (Figure 3.37B), the surrounding magnetic fields aid each other and so the overall inductance is increased. In this case the total inductance is L = L1 + L2 + 2M, and M represents the mutual inductance.

If the circuit is set up so that the current flow through the two coils is in opposite directions (C) the overall inductance will be diminished. In this case the total inductance will be L = L1 + L2 − 2M.

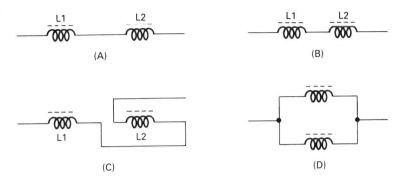

Figure 3.37. Inductors in series (no magnetic coupling) drawing (A); inductors in series with aiding magnetic coupling (B); in series with opposing magnetic coupling (C); inductors in parallel (D).

Inductors in Parallel

Coils can be wired in parallel (Figure 3.37D), and if there is no magnetic coupling between them, the total inductance is L = (L1 × L2)/(L1 + L2). Note the similarity of this formula to the one used for resistors in parallel.

Resistors, capacitors, and coils can be connected in a number of ways, a few of which are shown in Figure 3.38. In these fundamental circuits the resistor is the only electronic part that can be made so it is not frequency sensitive. The reactance of the coil increases with frequency; that of the capacitor decreases. Because of these three different electrical characteristics, combinations of resistors, capacitors, and coils can be used to perform a variety of different functions.

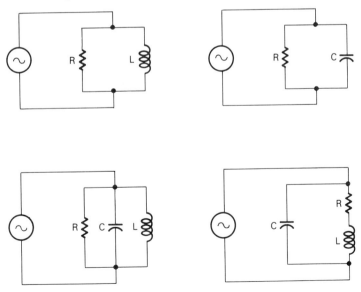

Figure 3.38. A few of the many possible R, C, and L combinations.

DIODES

As in the case of resistors, capacitors, and coils, there are various ways of interconnecting diodes. One method is to wire diodes in series (Figure 3.39A). A possible application is in voltage multiplier power supplies such as the half-wave tripler or full-wave quadrupler.

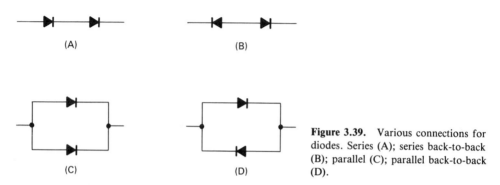

Figure 3.39. Various connections for diodes. Series (A); series back-to-back (B); parallel (C); parallel back-to-back (D).

Another method of connecting diodes is to wire them back-to-back (Figure 3.39B). Zener diodes are arranged this way across the secondary of a power transformer so as to supply regulated AC output. Diodes are sometimes placed back-to-back across a relay coil for protection on both halves of an AC input waveform when the coil is operated from an AC source.

Figure 3.39C shows a pair of diodes in parallel. For a full-wave power supply, diodes are essentially in parallel when connected to the secondary winding of a power transformer. Figure 3.39D illustrates diodes connected in reverse parallel. This has its counterpart in the back-to-back arrangement in B. This kind of circuit is used in a ratio detector, a demodulator for FM receivers.

TRANSISTORS

Transistors can be joined in a number of different ways, including series and parallel circuits (Figure 3.40). These find wide application in gating circuits, described in more detail in Logic Circuits, Chapter 7.

Decibel Reference Levels

A circuit diagram may contain some reference to voltage or power values in some form of decibels (dB).

The decibel is a comparison of two signal levels in terms of voltage, two current or two power levels. Where a comparison level is not supplied, a selected reference level can be used instead. Various abbreviations are used to indicate reference levels.

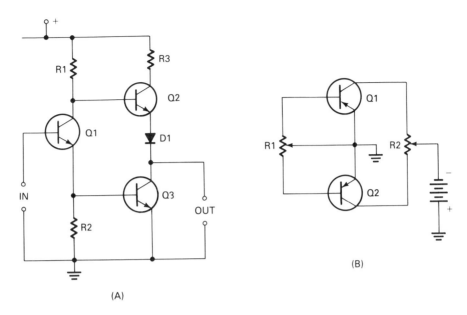

Figure 3.40.   Transistors in series (A); in parallel (B).

dB	Reference
dBj	1 millivolt
dBk	1 kilowatt
dBm	1 milliwatt, 600 ohms
dBs	Japanese designation for dBm
dBv	1 volt
dBw	1 watt
dBvg	voltage gain
dBrap	decibels above a reference acoustical power of 10^{-16} watt

When 1 watt is used as the comparison reference, 0 dBw is 1 watt; 10 dBw is 10 watts; 20 dBw is 100 watts.

Metric Prefixes and Symbols

A number of different metric prefixes are used in connection with circuit diagrams, and these are represented by alphabetic symbols.

	Prefix	Exponential Form	Symbol
	tera	10^{12}	T
	giga	10^{9}	G
	mega	10^{6}	M
	kilo	10^{3}	k
	hecto	10^{2}	h
	deca	10^{1}	dam
		10^{0}	
	deci	10^{-1}	dm
	centi	10^{-2}	cm
	milli	10^{-3}	mm
	micro	10^{-6}	μm
	nano	10^{-9}	n
	pico	10^{-12}	p
	femto	10^{-15}	f
	atto	10^{-18}	a

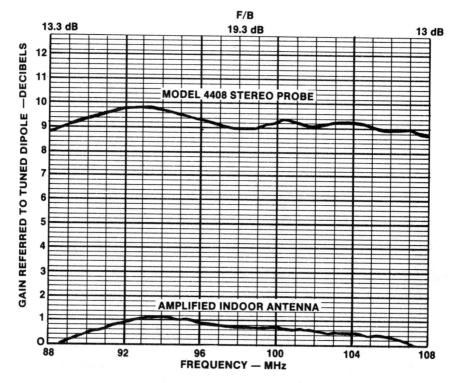

Figure 3.41. Gain of an antenna in dB at various frequencies (MHz)

There are certain conventions in using these abbreviations. Thus, a microsecond could be abbreviated as μs, and a thousandth of a microsecond as mμs, or a millimicrosecond. The preferred form is ns or nanosecond. The use of double prefixes is undesirable.

Symbols begin with lower case letters except for terameter, gigameter, and megameter. Thus, megameter and millimeter use the same letters in their symbols except that megameter is Mm and millimeter is mm. Gigameter has no competing symbol but is generally writted Gm.

Quite often both a reference level and a symbolic prefix will be used at the same time. The graph in Figure 3.41 is that of the gain of an antenna expressed in decibels (dB) and frequency in megahertz (MHz). Further information about graphs is supplied in Chapter 9.

COMPONENT COMBINATIONS

There are astonishingly few distinctive radio parts such as resistors, capacitors, coils, diodes, and transistors. And yet, with just this small number, hundreds of circuit diagrams can be developed and drawn. The reason for this is that each part isn't just a single item but has a substantial number of variations.

4

Subschematic Diagrams

There are three types of schematic diagrams (also called "circuit" diagrams): main schematics, partial schematics, and subschematics, also called "subcircuits." A main schematic is one that, using electronic symbols and possibly a partial block diagram, is that of an entire component. A main schematic of an FM receiver, for example, would be the diagram of that receiver from the input (the antenna), to the output (the speakers). A partial schematic, actually a circuit diagram within a circuit diagram, could be the complete RF (radio-frequency) amplifier of that receiver, or an IF (intermediate-frequency) amplifier, or a demodulator section, or an AF (audio-frequency) section, or the power supply. The sum of all these partial schematics is the entire receiver (Figure 4.1). Finally, a subschematic is some portion of a partial schematic. The subschematic could be used to analyze the functioning of some section of an RF amplifier, or AF amplifier.

Of these three types, the main schematic is the most elaborate or the largest; the partial schematic is smaller; the subschematic is the smallest. The subschematic may possibly consist of nothing more than a transistor, a resistor, and a capacitor.

CIRCUIT ANALYSIS

The circuit in Figure 4.2 is that of a complete power supply with vertical lines showing the division of this circuit into five subsections. The first subsection (A) consists

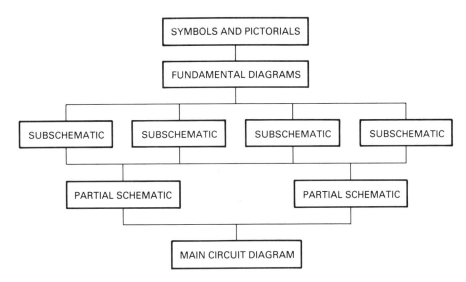

Figure 4.1. Steps in the formation of a main circuit diagram.

of the power transformer input and includes the fuse F1 and switch SW1. This is followed by the rectifier section (B) and shows the use of a pair of diodes in a full-wave arrangement. The output of the rectifier is brought into a filter (C), and this subsection consists of a pair of capacitors C1 and C2 plus a fixed-iron core choke, L1. Section (D) is a DC voltage regulator that includes a Zener diode, D3, and a transistor Q1. The output (E) is a voltage divider resistor, R5. The divider supplies various amounts of DC voltage.

Any sequential subcircuits, taken together, form a partial circuit. Subcircuits (A) and (B) for example, could be one, and (C) (D) and (E) another. Although a subcircuit usually consists of a combination of different parts, such as resistors, capacitors, and inductors, it is possible for it to contain just one of these. Subcircuit E, for example, uses just a single tapped resistor.

To simplify the circuit diagram, a subcircuit within that diagram may sometimes be shown in block diagram form. Thus, the voltage regulator section (D) contains a differential amplifier. This amplifier is a subcircuit in itself, and so it is possible to have a subcircuit within a subcircuit. Quite often a circuit diagram will use blocks to represent integrated circuits (ICs) or triangles to show the use of operational amplifiers (op amps).

POWER SUPPLIES

The DC voltages required by a circuit can be furnished by a battery or by a power supply connected to the AC power line.

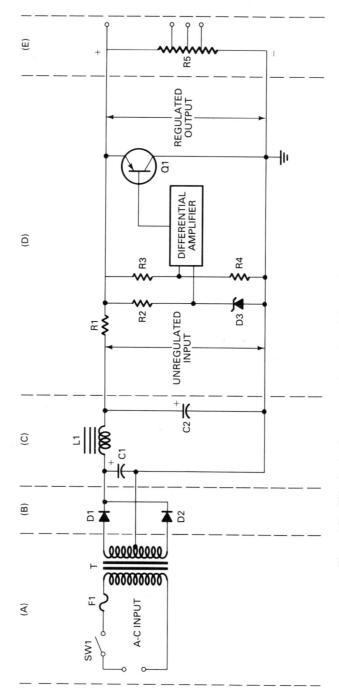

Figure 4.2. This circuit of a power supply consists of five subcircuits. Subcircuits can be combined to form partial circuits. Each subcircuit contains one or more fundamental circuits.

RECTIFIERS

The rectifier is a diode, and for industrial applications may be either a vacuum or a gas-tube type. For in-home entertainment systems such as receivers and TV sets, the rectifier is a solid-state semiconductor made of silicon.

Half-Wave Rectifier

The input to the rectifier can be the secondary winding of a power transformer, either a step-up, step-down, or a one-to-one type, or else the rectifier subcircuit can be connected directly to an AC power outlet.

The subcircuit in Figure 4.3, known as a half-wave rectifier, is the simplest and possibly the most widely used and consists of a single crystalline semiconductor. Current flows through the load resistor R only when the AC input makes the anode of the crystal positive. No current flows during the negative half of the input waveform, and so the output isn't continuous but is made up of a series of half-wave pulses having the same frequency as that of the AC input, usually 50 or 60 Hz.

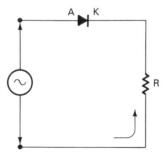

Figure 4.3. Half-wave rectifier.

Full-Wave Rectifier

The full-wave rectifier (Figure 4.4) works on both halves of the AC input cycle, and so its pulsed DC output has twice its frequency. For a frequency input of 60 Hz, the DC output ripple is 120 Hz. This type of DC output characteristic of this rectifier subcircuit, and others as well, can be changed into smooth, nonvarying DC by a filter and in some instances by a voltage regulator as well. One of the disadvantages of the full-wave rectifier is that its power transformer must be center-tapped.

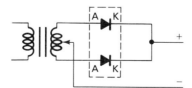

Figure 4.4. Full-wave rectifier. The two diodes may be enclosed in a package, as indicated by the dashed lines, but may be shown as separate units.

Full-Wave Bridge Rectifier

This rectifier subcircuit (Figure 4.5) uses four rectifier diodes, but has two important advantages. The power transformer does not require a center tap. Further, the diodes form series pairs. Thus, in Figure 4.5, D1 and D4 are in series and so are D2 and D3. Consequently, the inverse peak voltage, the voltage across the diodes during the negative half of the input cycle, is divided across the two diodes, thus putting each of the diodes under less electrical pressure. The pulse output frequency is the same as that of the two-diode full-wave type. Current flow through the load takes place during both halves of the AC input cycle.

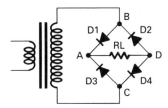

Figure 4.5.   Full-wave bridge rectifier.

Full-Wave Voltage Doubler

A pair of diodes can be used to supply full-wave rectification and at the same time can supply twice the DC output than that of the usual full-wave supply (Figure 4.6). This is obtained by using a pair of capacitors in series across the output. Each capacitor becomes charged in turn with the DC output equal to the sum of the voltages across the capacitors.

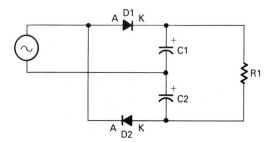

Figure 4.6.   Voltage doubler.

Cascade Voltage Doubler

Another type of voltage doubler is the cascade subcircuit shown in Figure 4.7. During one-half of the input cycle, capacitor C1, in series with diode D2, becomes charged. During the other half of that cycle, capacitor C1 charges capacitor C2 to twice the peak voltage of the AC input.

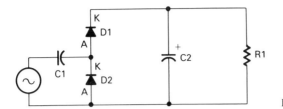

Figure 4.7. Cascade voltage doubler.

Bridge Voltage Doubler

The concept of using a pair of capacitors with each of them charged alternately by the AC power line to supply approximately twice the output voltage can also be applied to a bridge rectifier circuit (Figure 4.8A). In this illustration capacitor C1 is shunted across diode D4, and capacitor C2 is in parallel with diode D3. Since each of these diodes works on alternate halves of the input AC cycle, the capacitors become charged separately to the full AC line voltage. Since the capacitors are in series, their voltages are additive, and so the output across the load resistor is approximately twice that of the AC line voltage.

The concept of voltage multiplication isn't restricted to doublers and can be used to obtain much higher values of DC output using voltage tripler circuits, quadruplers (Figure 4.8B), and so on. These aren't too common since higher output voltages can be obtained by using a step-up power transformer. The transformer, however, adds weight and demands more room.

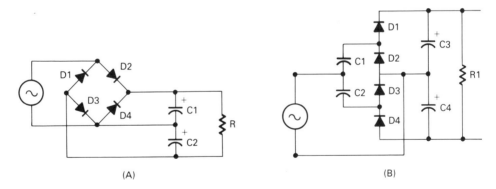

Figure 4.8. Bridge voltage doubler (A); quadrupler (B).

FILTERS

The pulsating DC output of the various rectifier subcircuits cannot be used as a voltage source without the addition of a filter. However, the power supply is just one of the many subcircuits in a component to which a filter can be applied.

Filters consist of resistance (R), inductance (L), and capacitance (C), but the simplest filters can be nothing more than a single capacitor or coil. The purpose of a filter is to permit the passage of some frequencies and to block others. The frequency or frequencies at which the blocking action takes place is known as the cutoff frequency, and there can be one or two such frequencies. The input and output connections of filters are terminated in source and load resistances or impedances equal in value to the impedance of the filter.

Filters can be used at any frequency common to electronic equipment. They are used in radio receivers in a variety of applications, in equalizers, in compact disc players, and in TV receivers. Filters are named for the work they do, as, for example, a band-pass filter, or for the subcircuit or partial circuit in which they function. Thus, a power-supply filter is one associated with a power supply. Filters named for the work they do would include the high-pass T filter, the high-pass pi filter, the low-pass T filter, the low-pass pi filter, the band pass filter, and the band stop filter. In some instances a filter is named for the type of waveform that is being used, as in the case of an analog filter or a digital filter. In some instances the name is rather dramatic, as, for example, the brickwall filter. The name of a filter can supply some indication as to its complexity, as in a 96th order filter. And filters can be named after their designers.

Power Supply Filters

A power supply filter can be used to follow any of the rectifier circuits shown earlier. The simplest would be a single capacitor shunted across the rectifier output (Figure 4.9A). Sometimes a series-tuned circuit is used as in B. This subcircuit consists of a capacitor C1 in series with a fixed iron-core coil L1. The combination is tuned to the pulse frequency of the rectifier output, and since it is a series circuit it will have minimum impedance at that frequency. This single frequency filter bypasses current variations and helps smooth the DC output.

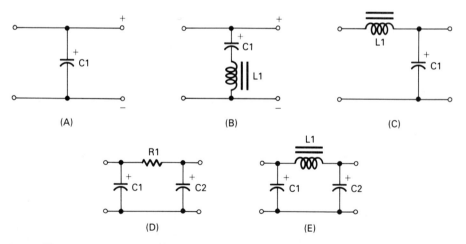

Figure 4.9. Power supply filters. Input at left is connected to rectifier. Output is to voltage regulator or load.

The filter in C is sometimes called a brute-force type. The choke coil, L1, has a high reactance to any current variations while capacitor C1 helps produce a more uniform DC output. The subcircuit in (D) consists of a pair of filter capacitors, one on either side of a resistor R. There are several disadvantages to using a resistor in the filter. It must be capable of passing all the current required by the load; there is not only a voltage drop across the resistor, thus decreasing the available output voltage, but there is a power loss in the resistor as well. An improved version of this circuit (E) uses an iron-core choke coil in place of the resistor.

The choke coils in these filters are generally rated at 5 henrys or more. The filter capacitors are electrolytics, ranging from 5 microfarads upward. As electrolytics they are polarized and are marked with plus signs in the circuit diagram. Quite often, though, these polarity indicators are omitted.

High-Pass Filter

A high-pass filter is one in which frequencies higher than the cutoff frequency f are passed, while frequencies lower than the cutoff frequency are attenuated (Figure 4.10). In this subcircuit the filter depends on the opposite kinds of behavior of the capacitor (C) and the inductor (L). With increasing frequency, the reactance of the capacitor decreases and that of the inductor increases. Consequently the filter tends to pass frequencies as they become increasingly higher. At DC (zero frequency) there is complete blockage because of capacitor C. For AC there is no sharp point at which frequencies are completely blocked. The resistor R is the load, and the voltage of the passed frequencies appears across it.

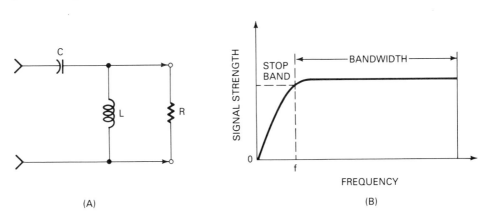

Figure 4.10. High-pass filter (A). Graph of operating characteristic (B).

To improve the performance of the filter, a number of C and L sections can follow the one shown in the figure, with the load resistor, R, appearing across the final one. The inductor L is shown here as an untuned, air-core type, but it can be equipped with a tunable iron core so as to help achieve a more precise value of inductance. Various values of L and C can be selected so as to achieve a specific cutoff frequency.

Low-Pass Filter

The low-pass filter (Figure 4.11A) works in a manner opposite that of the high-pass filter, and a comparison with the high-pass filter shows that its components have been transposed.

The low-pass filter will pass all frequencies below a cutoff frequency designated as f (Figure 4.11B). That is, the attenuation of all frequencies starting with zero, considering DC as the starting point, or zero frequency, up to the cutoff frequency will be minimum. This is just another way of saying that all these frequencies will be passed. But as shown in the graph, attenuation gradually increases with a rise in frequency. At some frequency the attenuation will be so large that we consider it as a rolloff point and indicate this by a symbol, f. The strength of higher frequency signals then gradually decreases. This curve is idealized. The cutoff is by no means as sharp, nor is the bandpass as uniform as shown.

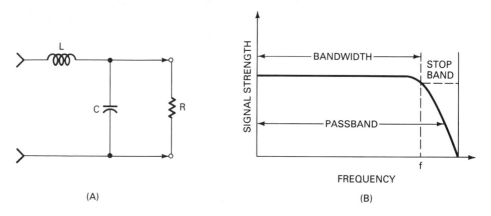

(A)   (B)

Figure 4.11.   Low-pass filter (A). The strength of the signal that is passed by the filter decreases as the frequency rises (B).

Band-Pass Filter

The band-pass filter is designed to pass a selected band of frequencies and to block the passage of frequencies below the selected pass band and above it. As shown in Figure 4.12A, this filter consists of a pair of tuned circuits, a series tuned consisting of L1 in series with C1, and L2 in parallel with C2. It is the combination of these two tuned circuits that forms the band-pass filter.

The series arm of this filter has its minimum impedance at the center frequency of the desired band. This takes place at its resonant frequency. The parallel-tuned arm has maximum impedance at its resonant frequency, and this is designed to take place at the center of the pass band. With this type of filter there are two cutoff points

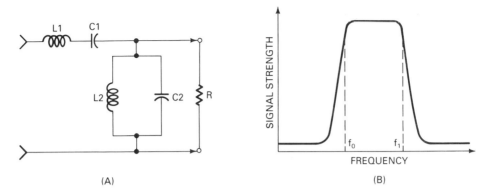

(A)   (B)

Figure 4.12. Band-pass filter (A). Operating curve (B).

(f_0 and f_1) one below the band of frequencies being passed, and the other above it (Figure 4.12B).

Cutoff Points

The cutoff points of low-pass, high-pass, and band-pass filters does not mean a sudden stoppage of the oncoming signal. How sharp the cutoff is depends on the number of elements in the filter, whether it is a single filter unit or one that has a number of repeated filters. Quite often, a gradual slope following the cutoff points is all that is necessary.

Band-Elimination Filter

The band-elimination filter (Figure 4.13A) is the opposite of the band-pass filter, and by comparing the two drawings you can see that the series and parallel tuned elements have been transposed. C1 and L1 have a maximum impedance at a specific

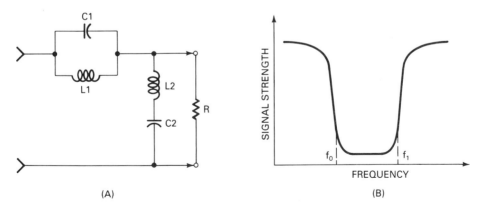

(A)   (B)

Figure 4.13. Band-elimination filter (A); operating curve (B).

frequency, or, as in this case, optimum impedance for a band of frequencies. On either side of that band the impedance decreases. The action of the series arm, L2 and C2, is exactly opposite. The frequency stopping action of C1 and L1 can be made very sharp (B) so that a limited or a wide band of frequencies is opposed.

T-Type Low-Pass Filter

It is difficult to produce a sharp frequency cutoff with single section filters. Cutoff can be improved in the case of the low-pass filter of Figure 4.11 by adding another coil in series with the original. Every circuit consisting of L and C is tuned, and so to maintain the original reactance of the coil it can consist of a unit divided into two sections, shown as L/2 in Figure 4.14. It is now called a T-filter because of its resemblance to that letter.

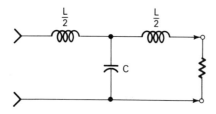

Figure 4.14.   T-type low-pass filter.

Pi-Type Low-Pass Filter

Two T-type low-pass filters can be combined (Figure 4.15), and the two inductors can be combined. Because of its appearance the filter is known as a pi (π) type.

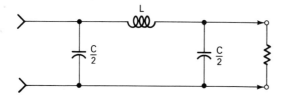

Figure 4.15.   Pi-type low-pass filter.

T-Type High-Pass Filter

The high-pass filter of Figure 4.10 can be changed into a T-type (Figure 4.16A) by inserting another series capacitor into the line. A multisection T-type, made by joining two such units, would have a pair of serially connected capacitors.

Pi-Type High-Pass Filter

Adding another shunt inductor, as in Figure 4.16B, gives us a single-section, pi-type high-pass filter. A two-section unit can be made by joining two single units. If the two inductors are identical, they can be replaced by a single unit having half the value of either.

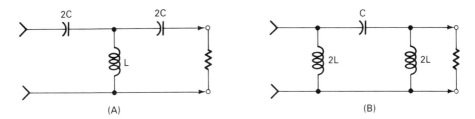

Figure 4.16. T-type filter (A) and pi-type high-pass filter (B).

Tuned Circuit Filters

A tuned circuit consisting of a coil shunted by a capacitor has its maximum impedance at its resonant frequency. A tuned circuit made up of a coil in series with a capacitor is exactly the opposite—it has a minimum impedance at its resonant frequency. It is possible to take advantage of these characteristics in the formation of filter subcircuits (Figure 4.17).

The illustration shows four possible arrangements. In Figure 4.17A inductor L1 and capacitor C1 are parallel tuned, and the subcircuit is a pi type. Figure 4.17B uses a series-tuned arrangement consisting of L2 and C2, and the filter is a T-type. Figure 4.17C is similar to A as far as the tuned section is concerned, but coils replace

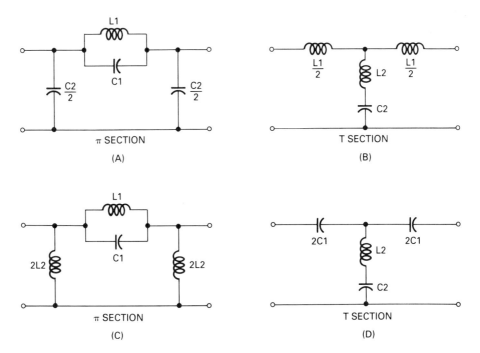

Figure 4.17. Tuned circuit filters.

the capacitors used previously. Capacitors have a reactance that decreases with frequency; coils work in an opposite manner and have a reactance that increases with
frequency. Figure 4.17D resembles B, except that a pair of capacitors are used to
replace the coils. This last subcircuit is a T-type filter. Tuned circuit filters are used
when a sharper cutoff is wanted.

Butterworth Filters

The characteristic curves that accompany the simple filters that have been illustrated
are idealized and are just a rough approximation of the results that can be obtained.
To get a better response more filter elements may be required. Figure 4.18 shows
three of these. Known as Butterworth filters, (A) is a low-pass, (B) is a highpass,
and (C) is a bandpass type.

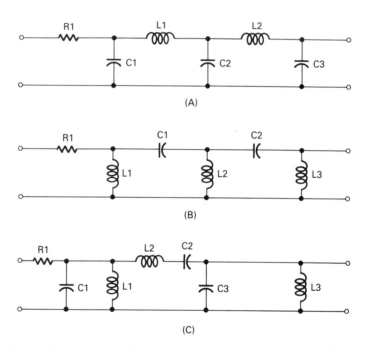

Figure 4.18.   Butterworth filters. (A) low-pass; (B) high-pass; (C) band-pass.

Chebyshev Filters

These filters are somewhat similar to the Butterworth units in the sense that they
are designed to supply better operating curves than ordinary filters (Figure 4.19). The
filter in (A) is a low-pass, (B) is a highpass, and (C) is a bandpass.

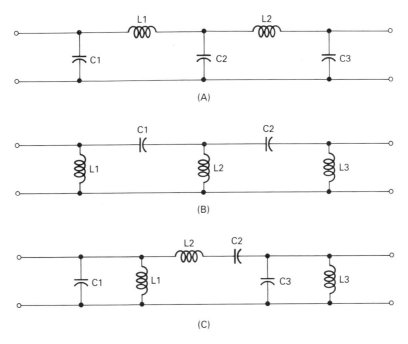

Figure 4.19. Chebyshev filters. (A) low-pass; (B) high-pass; (C) band-pass.

FILTER IMPEDANCES

The filter itself represents an impedance; that is, it has its own input and output impedances. These impedances, at each end of the filter, are known as image impedances. For maximum transfer of energy from the source to the load, the image impedances should be equal to the source and load impedances.

ZENER DIODE SUBCIRCUITS

Zener diodes are used in subcircuits for voltage regulation at the DC output of a low-voltage power supply, for the regulation of the AC input to a power supply, for the protection of meters, for interstage coupling, as a Zener-coupled amplifier, as a voltage reference, for biasing, as waveform clippers, and so on.

Low-Voltage Regulated Supply

A Zener diode can be placed across the DC output of a low-voltage power supply (Figure 4.20), and its function is to help maintain a constant amount of DC voltage for a load. In this circuit the input can range from 14 to 20 volts, and it is applied to the Zener through a low value of resistance (R1) possibly 10 ohms. The Zener is wired in series with the load.

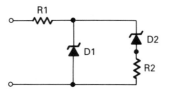

Figure 4.20.   Zener diode subcircuit for regulation of power supply output.

ATTENUATORS AND PADS

The purpose of an attenuator or a pad is to produce a previously determined amount of signal reduction, generally between one circuit and the next without disturbing the impedance relationship of those circuits. Attenuators use a combination of variable resistors; pads use fixed resistors only, but in either case no other parts such as coils and capacitors.

The L-attenuator (Figure 4.21A) is so-called since (with a little imagination) it seems to resemble an upside-down L. It has two variable resistors, mounted on a common shaft, so that the ohmic values of these resistors are changed at the same time.

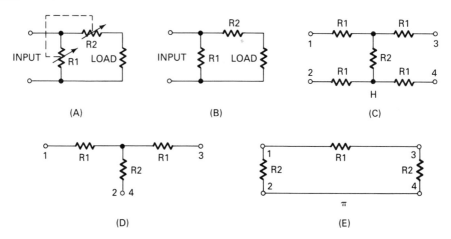

Figure 4.21.   L-attenuator (A) and pad (B); H pad (C); T pad (D) and pi (π) pad (E).

The attenuator has an input, a source voltage to which it is connected, and an output identified here as the load. Regardless of the setting of the variable resistors, the attenuator presents a constant resistance to the source voltage. Since maximum power is transferred when the resistance of the source is equal to the resistance of the load, the attenuator can be designed so that its input resistance always equals the source resistance regardless of the settings of R1 and R2. At the same time the attenuator drops the voltage to the amount required by the load and also matches the resistance of the load. The arrangement of the L-pad in (B) is the same as that in (A), except that fixed resistors are used. (C) is an H pad, (D) a T pad, and (E) a pi (π) pad.

NEUTRALIZATION SUBCIRCUITS

The transistor is capable of oscillating readily, and while this is a desirable characteristic for oscillator circuits, it is unwanted for amplifiers. To overcome the oscillating tendencies of a transistor, some part of the energy in the output circuit is fed back to the input. A prime requirement is that the returned signal should be out of phase with the input, something that can be done with neutralizing circuits, also called "negative feedback" circuits.

There are various subcircuits that can be used consisting of either a resistor or capacitor or a combination of the two (Figure 4.22). The simplest (A), consists of a feedback capacitor, C, connected from the bottom end of a tuned transformer in

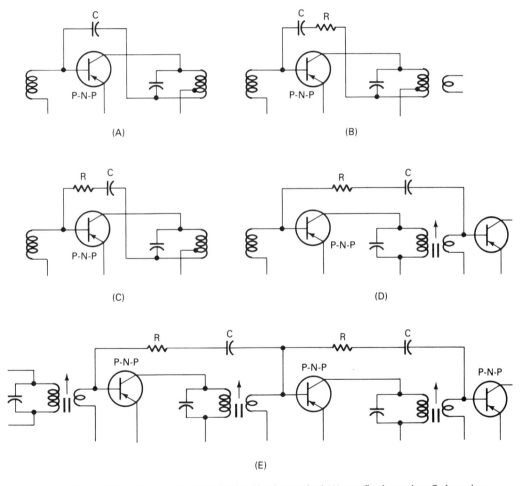

Figure 4.22. Negative feedback circuits. Simplest method (A) uses fixed capacitor. Series resistor can be inserted (B) for feedback control. R and C can be transposed (C). Negative feedback from base of following stage (D); feedback over several stages (E).

the collector circuit to the base. Since negative feedback reduces the gain of this amplifier circuit it may be necessary to restrict or control the amount of feedback. This can be done by selecting the value of capacitance of C very carefully or by putting a resistor in series with it (B). The location of R and C can be interchanged (C). In all three of these circuits the coil is tapped with the lower section of that coil as part of the feedback circuit. A tapped coil can be avoided, as in (D). Here the feedback signal is taken from the base of the transistor of the following stage, still using an R-C arrangement. Feedback is often needed in more than one amplifier stage, especially in intermediate-frequency amplifiers. Figure 4.22E shows the use of negative feedback over several stages.

OSCILLATOR SUBCIRCUITS

A transistor circuit can be made to oscillate, that is, to work as a generator of signals. This is done by taking some of the signal energy in the output circuit of a transistor and feeding it back in phase to the input. This procedure is the opposite of that used in neutralization. There are various ways of doing this, resulting in oscillator circuits having different names, but the basic principle is the same. Oscillator circuits have various applications, but every superheterodyne receiver, whether AM, FM, TV, or satellite, has one, and sometimes more than one.

Figure 4.23 shows the basic subcircuit concept. L1 and L2 form a transformer with the winding of L1 in the collector output circuit. Some of the signal energy in L1 is magnetically coupled to L2 in the base input circuit. In this case the fed back signal is in phase with the input.

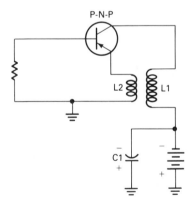

Figure 4.23. Oscillator circuit using electromagnetic feedback.

Hartley Oscillator

The Hartley oscillator (Figure 4.24) is a P-N-P transistor amplifier using positive feedback to force the subcircuit into oscillation. The frequency of oscillation is determined by the parallel tuned circuit L1, L2, and C1. Feedback, also called regenerative feedback, is achieved by using L1, L2 as an autotransformer with the amount

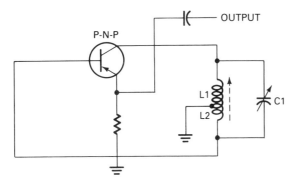

Figure 4.24. Hartley oscillator.

of feedback controlled by the position of the tap. C1 is a variable capacitor; L1, L2 is tunable, generally by a polyiron slug used as the core. The signal output of the oscillator can be fed into a mixer circuit to beat (heterodyne) with the incoming signal of a receiver. The result is an intermediate frequency (IF).

Colpitts Oscillator

This oscillator is essentially the same as the Hartley, but instead of using a tapped coil as an autotransformer it has a pair of series capacitors as a signal voltage divider. The tuned circuit (Figure 4.25A) consists of the coil, L1, and the total capacitance of C1 and C2 in series. Positive feedback is taken from the junction of the two capacitors back to the emitter. The amount of feedback is determined by the ratio of C2 to C1.

The Colpitts oscillator in Figure 4.25B uses a field-effect transistor instead of the P-N-P transistor in (A). And instead of using fixed capacitors, this circuit has

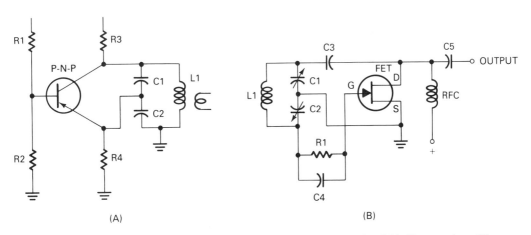

Figure 4.25. Colpitts oscillator using P-N-P transistor (A); with field-effect transistor (B). RFC is a radio-frequency choke. Its purpose is to help keep the oscillator's radio-frequency energy out of the power supply.

a pair of variable units mounted on a common shaft. The oscillator in (A) is fixed-tuned; that in (B) is variable. However, the capacitance ratio of the two variable capacitors remains fixed regardless of their rotation.

Phase Shift Oscillator

The output of a transistor amplifier whether it uses a P-N-P or N-P-N is 180° out of phase with its input. It can be made in phase by using a phase shifting network connected to the collector output.

    The subcircuit of the common-emitter amplifier in Figure 4.26 takes advantage of this concept by using a series of R-C networks consisting of R1-C1, R2-C2, and R3-C3. Each of these supplies a total phase shift of 60°, and so the sum of these is 180°. This, added to the normal phase shift of 180° by the transistor, produces 360°, or enough to make the output in phase with the input. In this arrangement all of the capacitors in the phase shifter must have the same value, and that is also true of all the resistors.

    In the phase shifter, R1 and C1 deliver the signal to R2 and C2 at the same time supplying a phase shift of 60°. There is a phase shift every time the signal is moved along from one R-C network to the next. The frequency of oscillation of this circuit is determined by the values of resistance and capacitance in the phase-shifting network.

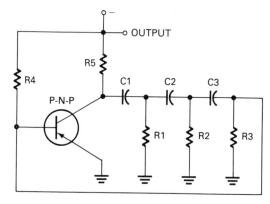

Figure 4.26. Phase shift oscillator.

Crystal Oscillator

The common problem associated with the oscillator subcircuits described previously is that they are subject to frequency drift. This can be overcome through the use of a piezoelectric crystal together with a bipolar transistor (either N-P-N or P-N-P) (Figure 4.27A) or a FET (field-effect transistor) (Figure 4.27B).

    In Figure 4.27A the crystal is in the input circuit of the N-P-N transistor and determines the oscillating frequency. In some crystal oscillator circuits, the collector output is also tuned. This tuned circuit can be adjusted to the fundamental frequency of the crystal, or to one of its harmonics if a higher operating frequency is wanted.

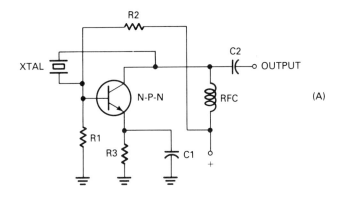

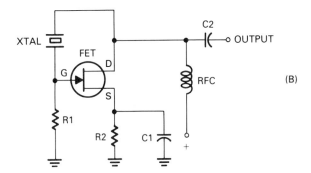

Figure 4.27. Crystal oscillator using N-P-N transistor (A); using field-effect transistor (B).

Pierce Oscillator

A crystal can be used with the oscillators described previously, in which case the circuit takes its name from the oscillator being used. Thus we have a Colpitts crystal oscillator, a Hartley, and so on. The oscillator in Figure 4.27B is a Pierce using a field-effect transistor with the crystal positioned between the gate and the drain, or, in the bipolar transistor, between the collector and base (Figure 4.27A). The advantage of the Pierce crystal oscillator is that it eliminates the need for a tuned L-C circuit in the output with the crystal acting as its own tuned circuit.

Multivibrators

All of the oscillators described previously produce a sine-wave output, but non-sinusoidal types are also essential. One of the most widely used of these is the multivibrator (MV), and there are a number of different types that come under this classification.

Known as an astable or free-running MV, the unit in Figure 4.28 uses a pair of P-N-P transistors wired in such a way that the output of the first is fed into the

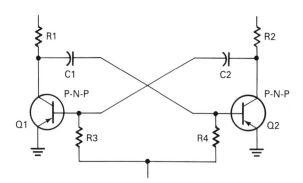

Figure 4.28. Multivibrator oscillator.

input of the second. Similarly the output of the second transistor is brought back to the input of the first. This crossfeeding is done by capacitors C1 and C2.

Coupling capacitor C1, in association with resistor R4, causes transistor Q2 to be cut off with the absence of current flow through this transistor dependent on the time constant of the R-C network. When C1 is discharged through R4, transistor Q2 goes into full conduction. Resistors R3 and R4 supply bias for their respective transistors. Bias voltage for Q1 is obtained by the discharge of C2 through R3. Similarly, bias for Q2 is supplied by the discharge of C1 through R4. As a result, each transistor alternately conducts and then is cut off.

Each transistor conducts, alternately, for a half cycle, with the output waveform a square wave. The frequency of the MV is almost solely determined by the values of C1,R4 and C2,R3, and can extend from a cycle or two up into the megahertz region.

BIASING METHODS

The biasing of a transistor determines its mode of operation. All that a triode bipolar transistor needs is forward bias for its base-emitter circuit and reverse bias for its collector-emitter circuit.

Figure 4.29 shows battery biasing for a P-N-P (A) and an N-P-N transistor (B). In the P-N-P transistor, battery B1 supplies bias by making the base negative with respect to the emitter. For the N-P-N the battery is reversed. An input signal can be applied across R1 while the output is taken from across R2.

One of the batteries can be eliminated as in (C), an arrangement using an N-P-N. Note that the collector is still made positive with respect to the emitter while the base is also made positive. Generally, a single battery is used to supply both collector and bias voltage with the required voltages obtained by using a series resistor network across a battery (D).

In transistors the base is biased either negative or positive with respect to the emitter, depending on whether the transistor is a P-N-P or N-P-N type. An increase in bias, regardless of the transistor, means an increase in collector current. In bipolar

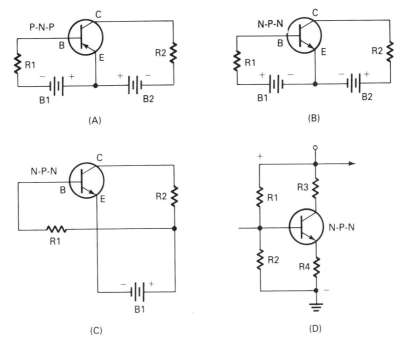

Figure 4.29. Biasing methods for P-N-P transistor (A); for N-P-N transistor (B). A single battery can be used as in (C). A series resistor voltage divider (R1 and R2) is more commonly used (D). These resistors are shunted across the battery supply. Bias voltage is also produced by the voltage drop across emitter resistor, R4.

transistors, bias current is the current moving between the emitter and the base. In a field-effect transistor, the bias is either the voltage between the source and the gate or the source and the drain (Figure 4.30A).

Figure 4.30B shows various techniques for biasing single gate MOSFETs. The one at the left is a self-bias circuit, that in the center is a fixed bias supply, while that at the right uses both techniques: self- and fixed bias. Generally, the fixed bias supply is not used, while the subcircuit employing both self- and fixed bias is preferred. Fixed bias is that furnished by a power supply and does not vary. Self-bias is produced by current flow through the transistor and for a bipolar transistor depends on the voltage drop due to emitter current flowing through the emitter resistor. For the FET it is due to the voltage drop resulting from current flow through the source resistor.

In either case, whether an emitter or source resistor is used, it may be bypassed by a capacitor. The current flow through the transistor varies in accordance with the input signal. This results in a varying DC voltage across the resistor. The shunting bypass capacitor helps smooth this voltage (Figure 4.31).

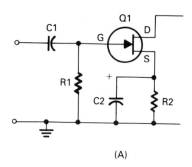

(A)

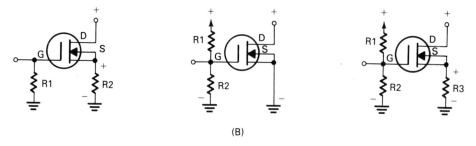

(B)

Figure 4.30. Biasing for FET (A); biasing for single gate MOSFETS (B).

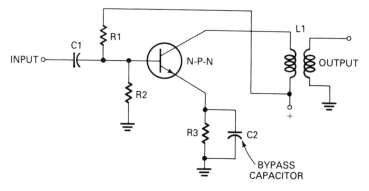

Figure 4.31. R3 is emitter bias resistor shunted by bypass capacitor C2.

GAIN CONTROLS

Gain controls, also called volume controls, are used on all radio receivers for adjusting the sound level output (Figure 4.32). Figure 4.32A uses an N-P-N transistor with variable resistor R1, the volume control across which the output signal is produced. Capacitor C1 permits passage of the signal to the potentiometer but blocks DC from the voltage supply. Since the amplified signal will be delivered to the base-emitter

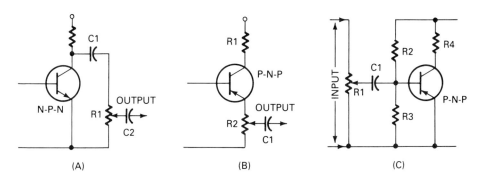

Figure 4.32.   Gain control subcircuits.

circuit of a following amplifier, another blocking capacitor, C2, is required. The base-emitter circuit of the next amplifier stage obtains its bias from a voltage source.

Figure 4.32B uses a P-N-P transistor with the load resistor, R2, in the emitter circuit. The variable resistor not only works as a load but as the volume control as well. This potentiometer must be designed to carry the transistor current.

In (A) and (B) the volume control is positioned on the output side of the transistor; in (C) it is located at the input. The volume control is isolated from the bias voltage of the following stage by capacitor C1. Resistors R2 and R3 are in series and are shunted across the voltage supply. These resistors supply bias for the P-N-P transistor. R4 is the load resistor for this transistor.

LOUDNESS CONTROL

A loudness control (Figure 4.33) is sometimes confused with a volume control since volume and loudness are closely related. Further, the loudness control does affect volume. The circuit is not found in moderately priced radios but is a subcircuit contained in a high-fidelity receiver.

The loudness control should only be turned on when listening at low volume levels. Human hearing isn't linear and discriminates against bass and treble tones under these conditions. With a fixed loudness control, bass and treble tones are automatically augmented. The variable loudness control is more desirable since bass and treble tones can be compensated to suit an individual's hearing capabilities, plus acoustic listening conditions.

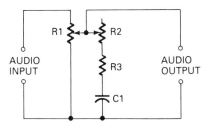

Figure 4.33.   Variable loudness control.

PRE-EMPHASIS AND DE-EMPHASIS

In FM transmitters audio frequencies are deliberately augmented toward the higher end of the audio range, a process known as "pre-emphasis." In the receiver a reverse process, "de-emphasis," is used to restore the audio signal to its proper level (Figure 4.34). Consisting of a capacitor C1 and a resistor R1, the combination has a time constant of 75 microseconds, obtained by multiplying R by C. With increasing audio frequencies, the reactance of C1 has a greater bypassing effect, thus gradually lowering the amplitude of the audio signal.

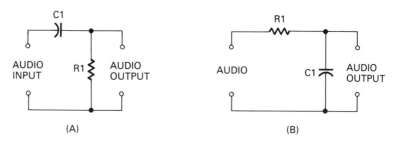

Figure 4.34.   Pre-emphasis network (A); de-emphasis (B).

AMPLIFIERS

Amplifiers were mentioned briefly in an examination of their fundamental circuits (Figure 3.15 and 3.16). These are now shown in more detail in Figure 4.35. Probably the most popular of these is the common emitter arrangement in Figure 4.35A. Fixed base bias is supplied by a pair of series resistors R1 and R2 shunted across the DC power supply. Self-bias is supplied by a resistor, R3, connected from the emitter to ground. Current flow through this resistor makes the emitter positive with respect to ground. The electrolytic capacitor, C2, across this resistor bypasses any signal currents, and so the voltage drop across R3 is steady DC. Fixed bias, as its name implies, is a constant value of bias voltage and is independent of the strength of the signal. Self-bias is produced by the flow of current through a transistor, with the amount of current dependent on signal strength. This current, flowing through a resistor, produces a bias voltage. This bias, working in conjunction with fixed bias, determines the operating point of a transistor. While self-bias is the term most popularly used, it is also known as "automatic" bias, and in the case of vacuum tubes is sometimes called "automatic grid" bias. All three configurations use both types of bias.

Of the three basic types of amplifiers, common emitter, common base, and common collector, the common emitter is the only one that has signal phase inversion. The output signal of this amplifier is 180° out of phase with its input, and for this reason lends itself well to the use of negative feedback. While the three illustrations all use an N-P-N transistor, a P-N-P type can be substituted with a transposition of the DC supply voltage.

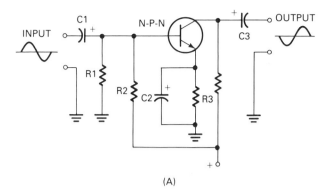

(A)

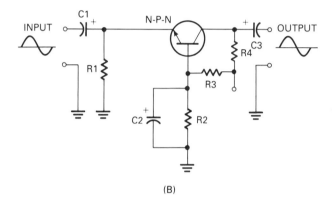

(B)

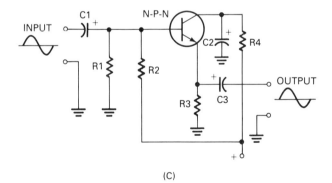

(C)

Figure 4.35. The three main amplifier configurations. Common emitter (A); common base (B); common collector (C).

The inputs and outputs of the three circuits is via an electrolytic capacitor, and since these capacitors are polarized, must be identified as such in the diagram. Further, their plus terminals must be connected to the plus side of any voltage dropping resistor. The names of these three amplifier circuits are taken from the way in which they are used. In the common base circuit, the input is via the emitter; the output is from the collector with the base common to both. In the common emitter, the input is to the base; the output from the collector, with the emitter as the common element. In the common collector, the input is to the base; the output is from the emitter, and the collector is the element common to these two. All three circuits are active types and consist of R-C networks used in combination with a transistor.

Differential Amplifier

The differential amplifier (Figure 4.36), unlike other amplifiers, has two inputs, not one, and also has two outputs. This amplifier is so-called since it functions on the difference between the amplitudes of the two inputs. Thus, if two identical signals are applied to the inputs, the difference between them (subtracting the strength of one signal from the other) will be zero, consequently the output will be zero. However, as the difference between the two signals increases, the output becomes greater accordingly.

Amplifiers of this type, often supplied in integrated circuits, have numerous applications. They are used in mixer circuits, frequency multipliers, demodulators, and so on. The signal output of this circuit is referred to as "differential voltage gain."

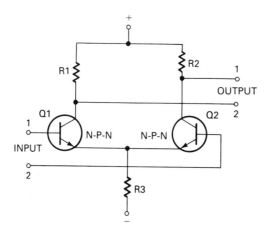

Figure 4.36. Differential amplifier.

Darlington Amplifier

The Darlington amplifier (Figure 4.37) consists of a pair of bipolar transistors, with both as P-N-P units or N-P-N. As shown in the figure, the two collectors are connected, while the emitter of the first transistor is connected to the base of the second. Sometimes known as a double emitter follower or a Darlington pair, the two transistors belong to the family of direct-coupled amplifiers. The output is taken from the common collector connection. While separate transistors can be used, the two are often combined as part of an integrated circuit.

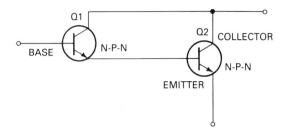

Figure 4.37. Darlington amplifier.

Figure 4.38 shows a more detailed arrangement of a circuit using a Darlington pair. The first transistor, Q1, uses both fixed and self-bias; the second transistor, Q2, self-bias only. The input impedance is high; the output impedance is low, with the signal output taken from the emitter of the second transistor. The collectors of both transistors are at RF ground potential obtained through a large capacitance electrolytic capacitor, C2. The gain of the circuit can be controlled by variable resistor R1 connected between the base of the first transistor and ground. The first transistor, Q1, works as an emitter follower.

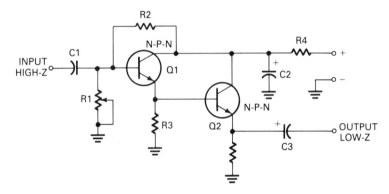

Figure 4.38. Details of Darlington amplifier circuit.

Emitter Follower

Typically, the output of a transistor amplifier is from the collector of one transistor to the base of a following transistor. However, in some circuits the emitter is used as the output, as in the Darlington pair. A circuit in which the emitter of a transistor is used for supplying the output signal is referred to as an emitter follower (Figure 4.39), and the connection between the output emitter and the base of the following stage is known as emitter coupling. When used alone, the emitter follower has a voltage gain of less than 1; that is, the circuit does not supply signal gain. However, as in the Darlington pair and in Figure 4.39 the emitter follower drives a transistor in a standard amplifier circuit and so the output does show substantial gain. Emitter followers are used for the purpose of impedance matching.

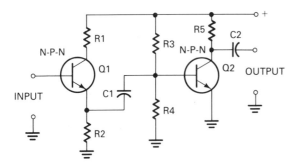

Figure 4.39. Emitter follower.

The emitter follower is not a direct-coupled unit, because a coupling capacitor, C1, is used to connect the emitter output to the base input of the following stage. In an emitter follower the output signal is in phase with the input. Either N-P-N or P-N-P transistors can be used in this circuit.

Direct-Coupled Amplifier

The Darlington is an example of an amplifier using direct coupling between an emitter driver and a following amplifier with base input. However, a commonly used circuit is one in which the collector of one stage supplies the signal directly to the base of a following transistor (Figure 4.40). In this circuit two different bipolar transistors are used, an N-P-N followed by a P-N-P. The input and output impedances are fairly low, in the order of about 1K ohms. The circuit uses an R-C decoupling filter consisting of R7 and C4, necessary since the DC voltage supply is used by both stages. The filter capacitor has the effect of making the output of the power supply a very low impedance, useful in keeping the signal from working its way through the power supply from one stage to the next. The absence of bypass capacitors across the emitter resistors, R3 and R5, supplies a small amount of negative feedback and helps stabilize the amplifier.

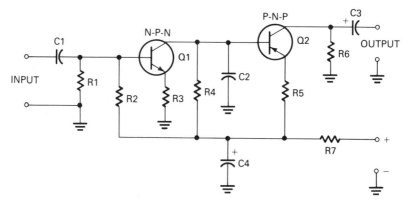

Figure 4.40. Direct-coupled amplifier.

Controlled Feedback Amplifier

Two types of signal feedback are used: negative and positive. Negative feedback reduces the gain of an amplifier, adds circuit stability, and improves frequency response. Positive feedback increases gain and narrows the bandpass, but if extended can make an amplifier fall into oscillation.

Ordinarily, the type of feedback and the amount is preset at the time a component is manufactured. However it is possible to control feedback through the use of an R-C arrangement (Figure 4.41). In this amplifier the signal appearing at the collector is 180° out of phase with the input and is returned to the input via R1-C1. The capacitor does introduce some phase shift of its own so the signal that is fed back isn't precisely out of phase with the input. The potentiometer controls the amount of feedback.

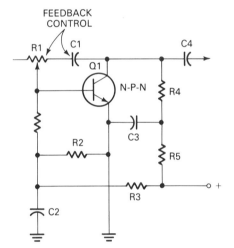

Figure 4.41. Controlled feedback circuit.

INTERSTAGE COUPLING

There are a number of ways in which one circuit can be coupled to the next: impedance, capacitive, transformer, and direct coupling. Impedance coupling, used at one time and now rarely seen, consists of an inductor across which the signal voltage is developed. Capacitive coupling uses a capacitor to link the signal, while transformer coupling uses a pair of coils. With direct coupling the output of one stage is led directly, without the use of coils or capacitors, into a following stage. Transformer coupling is used in RF amplifiers; capacitive or direct coupling in AF amplifiers. Transformer coupling was widely used at one time in audio amplifiers, but presently most such amplifiers use either direct or capacitive coupling.

CASCODE AMPLIFIER

Cascode amplifiers are used in RF amplifiers and also in audio preamplifiers. In RF applications they are characterized by their ability to increase high-frequency sensitivity, and in audio they are desirable because of their low level of distortion. The circuit arrangement (Figure 4.42) consists of a pair of N-P-N transistors with the collector output of a common-emitter driving the emitter of Q1.

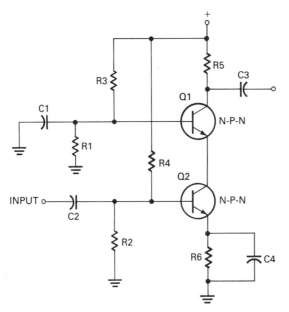

Figure 4.42. Cascode circuit.

COMPRESSION AMPLIFIER

A compression amplifier is one that works inversely with the amplitude of the input signal. The larger the amplitude of the input, the lower the gain of this transistor amplifier. Signal compression is obtained by using a diode rectifier so that its DC output modifies the bias to reduce the gain of the amplifying transistor (Figure 4.43).

CROSSOVER NETWORKS

A crossover network is positioned between the output of an audio power amplifier and the speakers to which it delivers its signals. Crossovers are filters for separating the audio band into two and sometimes three groups of frequencies: bass, midrange, and treble.

Crossover filters can be passive, consisting of capacitors only, or some combination of inductors and capacitors. There are also active types, better known as elec-

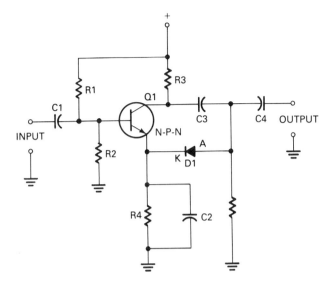

Figure 4.43. Compression amplifier.

tronic crossovers. When passive units are used they are generally housed in the same enclosure as their speaker systems while electronic crossovers are separate components.

The simplest crossover consists of a single capacitor put in series with one lead connected to a tweeter/midrange speaker while the woofer is shunted directly across the complete output (Figure 4.44A). The capacitor has a reactance that varies inversely with frequency and so it opposes low audio frequencies but has a decreasing opposition for tones in the midrange and least for treble tones. In this arrangement all tones, without limitation, are supplied to the woofer. At some frequency the reactance of the capacitor will be equal to the impedance of the tweeter, ordinarily 8 ohms.

To avoid supplying the woofer with treble tones, an iron-core inductor can be used in series with the voice coil of the woofer. The coil has a reactance which increases directly with frequency, and so, as in Figure 4.44B, treble tones are kept out of the woofer. The capacitor in series with the voice coil of the tweeter/midrange speaker has a much higher reactance to bass tones than to midrange or treble frequencies.

In another subcircuit (C) the woofer/midrange has its voice coil placed directly across the output of the audio power amplifier; consequently, signals of all frequencies are supplied to this driver. However, the tweeter has a filter consisting of a capacitor C1 and a coil L1. The reactance of the capacitor is low for treble tones and much higher for those in the bass and midrange. The coil has a high reactance for treble frequencies but permits the lower frequencies of the bass and midrange to be bypassed. These two parts, working together, manage to keep most of the bass and midrange frequencies out of the tweeter. The crossover network is followed by a potentiometer to control the amount of signal strength delivered to the tweeter. Some listeners find a high level of treble tones offensive. Further, lowering treble tone response this way makes bass and midrange tones seem stronger by comparison.

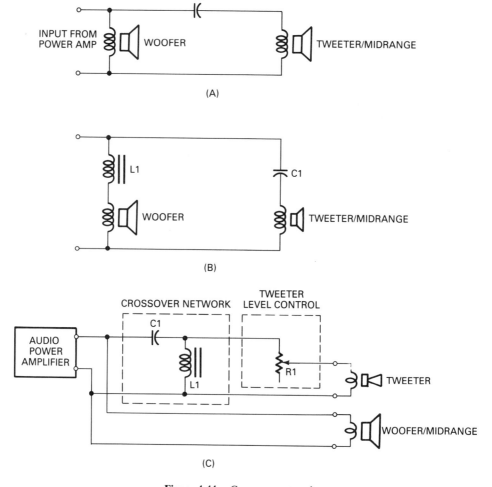

Figure 4.44. Crossover networks.

INTEGRATING AND DIFFERENTIATING NETWORKS

An integrating circuit (Figure 4.45A) consists of a resistor and capacitor and may also include an operational amplifier (op amp) described in more detail in the following chapter. These circuits are used in TV receivers to distinguish vertical from horizontal synchronizing (sync) pulses. While the figure shows just a single R-C, the integrator can be cascaded and in that case is followed by two or more similar R-C arrangements.

Integrating networks are inserted between the vertical sync amplifier and the following vertical sweep oscillator.

Figure 4.45. Integrating network (A); differentiator (B).

A differentiating network (Figure 4.45B) resembles an integrator but with the capacitor and resistor transposed. The function of this circuit is to separate horizontal sync pulses from the combined sync signal in a TV receiver. While this subcircuit has a resemblance to an integrator, its RC time constant is much shorter. The output of the differentiating network is taken from across the resistor instead of the capacitor as in the case of the integrator.

DIODE SWITCHING

A common method for switching circuits is to use some kind of mechanical device, and while this could be satisfactory for some applications, such as turning power on and off for some component, there are many instances in which such switching would be unsuitable. Mechanical switching requires manual control, and this often means it must be front-panel mounted. Such switching has two other features either one of which, or both, could be regarded as a fault. Mechanical switching is inherently slow, and while it would be satisfactory for turning AC input power on or off, there are various circuits that must be turned on or off very quickly under certain operating conditions. Further, mechanical switching is much more prone to breakdown than a purely electronic component.

A diode (Figure 4.46A) can be used as an electronic switch. Thus, a diode can be biased to produce a condition of cutoff. An input pulse equal in strength to the amount of bias, or greater than the amount of bias, will take the diode out of cutoff, permitting it to conduct. The amount of diode bias can be arranged so that the diode will conduct for small, moderate, or large amplitudes of input signal, and so the diode can be made to be signal discriminatory. A diode at cutoff is equivalent to an open switch, and can be considered a closed switch when it conducts.

The circuit in Figure 4.46A is a FET oscillator. With the application of a suitable DC voltage through R1, diode D1 will conduct, thus effectively grounding one end of C2, putting it in shunt with C1. This will change the operating frequency of the oscillator, decreasing it since more capacitance will be added. C2 can be switched in or out of the circuit by the diode.

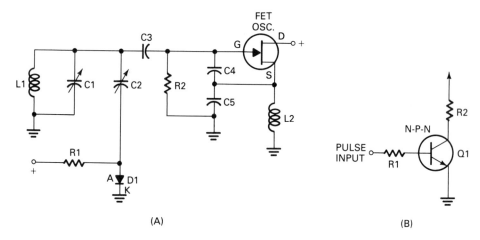

Figure 4.46.   Diode switch (A); transistor switch (B).

TRANSISTOR SWITCHING

As in the case of a diode, a transistor can also be used as an electronic switch (Figure 4.46B). Using a bipolar transistor the base can work as the control element for current flow between the emitter and collector. The base can be reverse-biased so as to block current flow, driving the transistor into its cutoff region. To reverse-bias a P-N-P unit, the base is made positive with respect to the emitter. For an N-P-N an opposite procedure is followed, and the base is made negative.

The input to a transistor used as a switch can be a pulse waveform applied between base and emitter. If the pulse amplitude is made large enough, the transistor will not only be taken out of cutoff, but will reach maximum current between emitter and base. For a P-N-P transistor, the input pulses must be negative going; for an N-P-N they must be positive going.

5

PCs, ICs, and Op Amps

At one time all wiring of a component such as a radio receiver or a television set was done on a point-to-point basis with individual wires connected from one component to another and with considerable manual soldering. The underside of a metal chassis looked like a wiring maze, required extensive hand labor in manufacturing, with time-consuming servicing not only a search for defective parts but for miswiring and cold solder joints. Wiring errors were not uncommon and not completely overcome by quality control. Printed circuit (PC) boards were developed not only to minimize these problems but to miniaturize components as well. While portable components, for example, preceded the use of printed circuits, they were not only heavy but bulky as well. In-home radio receivers were often housed in large consoles.

PRINTED CIRCUITS

A printed circuit (Figure 5.1) consists of a plastic laminate or other nonconductive material on which conductors are etched, sprayed, or painted, forming a predesigned pattern. Sometimes referred to as a "substrate," holes are drilled in this base for the insertion of components that are automatically connected when soldered into position. In some instances soldering is done on an individual point basis, but usually the entire board and its mounted components are soldered at one time by machine.

CONDUCTORS

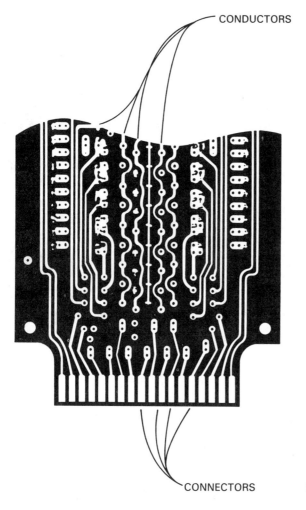

CONNECTORS   **Figure 5.1.** Printed-circuit board.

With the help of printed circuit boards (PC boards) and miniaturized parts, such as resistors, components of all types have been made much more compact, manufacturing costs have been substantially reduced, the overall weight has come down, and servicing and replacement have been made much easier.

PRINTED CIRCUIT BOARD PICTORIALS

Pictorials are often used with printed circuit board components and, for larger units, can be quite elaborate. However, some are quite simple (Figure 5.2A) such as this audio oscillator, shown together with its circuit diagram (B). In some instances the parts on the pictorial are identified by the same codes used on the schematic.

If the board and the parts are supplied as a package, possibly for a kit, construction is greatly simplified. As a further advantage it is an excellent way of learn-

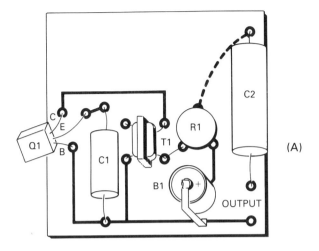

(A)

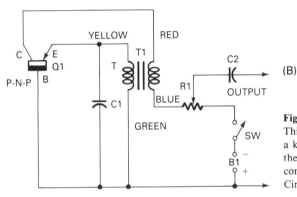

(B)

Figure 5.2. PC board pictorial (A). This type of drawing often accompanies a kit. The solid lines are connectors on the board. The dashed line is a wired connection to be made below the board. Circuit diagram (B).

ing the symbols and also how to read a circuit diagram. The placement of the parts is predetermined by the board's layout, resulting in little possibility for error. As a further advantage, the parts can be mounted and then soldered all at the same time. No drilling of the board is required.

PARTS PLACEMENT DIAGRAM

Parts placement diagrams are available to facilitate the exact positioning of electronics parts (Figure 5.3), not only for construction but for servicing as well. The diagram shown here is that of a simple setup, and while some parts placement diagrams are complex, they are still extremely helpful in locating individual components. Electronic parts may be mounted on one side of the board only, or both sides may be used. In that case an aboveboard and belowboard parts placement diagram may be needed.

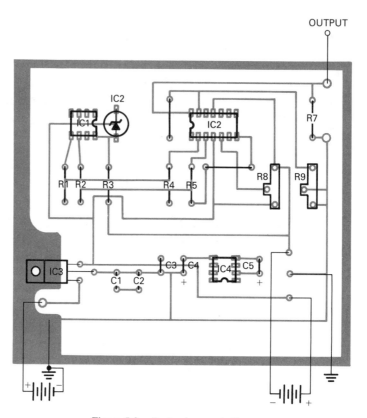

Figure 5.3. Parts placement diagram.

With some printed circuits, all the parts must be soldered in position. However, for those that use integrated circuits (ICs) sockets may be supplied and so the ICs are simply pushed into place. Since an IC may contain a number of prewired circuits, project construction and manufacturing of a component as well is greatly simplified. One advantage here is that this eliminates the need for soldering the connections to the IC. Not only can these be numerous, but soldering to an IC if not done properly can destroy the component.

PC BOARD SOLDERING

While printed circuit boards have numerous advantages, they can present some difficulties for the electronics hobbyist. The boards do not lend themselves easily to changes, and so circuit modifications may be difficult to do. The printed conductors may be very close to each other, hence carelessness in soldering can easily result in a short. Removing an incorrect or defective part can be troublesome.

Solder for PC boards is a tin-lead mixture containing rosin whose ability to work as a flux is due to abietic acid, but not all abietic acids used in solder are alike.

Some of these tend to spatter and when that happens high-resistance joints or short circuits can result. A servicing test is to try the solder first on a discarded PC board. Typical PC board solders are 50/50 lead/tin mixtures or one containing 63 percent tin and 37 percent lead, sometimes referred to as a "eutectic." Eutectic is characterized by the fact that it has the lowest melting point and so is desirable for PC boards.

PC BOARD ARRANGEMENTS

PC boards can be arranged in a number of ways. It can be a single board mounted horizontally, or a pair of boards, positioned horizontally parallel to each other. In some instances the boards are fairly small, carry discrete circuitry, and are all placed in a horizontal plane. In other arrangements boards can be put at right angles to each other. These setups depend on the available space in the cabinet.

With a few exceptions, most circuits today are solid state, and so the problem of how to mount and position vacuum tubes and allow space for them is eliminated. Semiconductor diodes and transistors, with the exception of power transistors, are tiny, lightweight, do not require sockets, and are self-supporting.

Motherboard

In some instances it may be necessary to join two or more PC boards to a main board, with that board containing the principal circuit system known as the motherboard. The motherboard contains edge connectors to facilitate the quick joining of supplementary boards.

Boards to be joined may be equipped with a polarizing slot, a notch that ensures correct mating (Figure 5.4A).

PC BOARD GROUND

At one time components such as receivers and amplifiers used a metal chassis, which not only supported the individual parts but acted as a convenient common connection, frequently called "ground" even though there might not have been any external wire to ground. The board on which wiring is placed is nonconductive and so an individual conductor is used as the common lead. In a circuit diagram this conductor is emphasized as a thicker line than the others, although this conductor in an actual PC board may have no more thickness than any other. Figure 5.4B is part of circuit board diagram.

PARTS FOR PC BOARDS

Resistors for PC boards can include those made from resistance paints, tape resistors made from a mixture of resin and carbon, subminiature carbon resistors, subminiature wirewound, deposited film, flexible film, and miniature variable. Wattage ratings

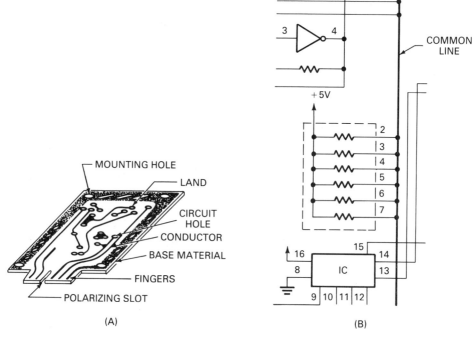

Figure 5.4. PC board with polarizing slot (A). Portion of PC board circuit diagram with common line emphasized (B).

are from about 1/5 watt to 1/2. While larger wattage rating resistors can be used, the problem is one of accommodating their larger size.

Capacitors include flat miniature ceramic, tubular, and disc ceramic types, metallized paper, mica, electrolytics, tantalum, and variable capacitors. Inductors include coils, both form wound and flat spiral, and transformers.

PUNCHBOARDS

A disadvantage of a printed circuit board is that once the pattern has been determined and produced, changes are rarely practical. This is an ideal arrangement when large numbers of boards having identical patterns are to be manufactured, but this benefit does not accrue for experimenters, hobbyists, or those who use kits for learning the behavior of a number of circuits. For such applications, punchboards are much more practical. Punchboards can be made of any insulating material including pressed board, Masonite, plastic, and so on, and are available in various dimensions and materials. The boards are prepunched with holes. The hole sizes selected for a particular board could have a diameter of 3/32 inch, but other holes sizes are available as well. Board thickness ranges from 1/16 to 3/32 inch. The holes are arranged in rows and columns, and a representative board could have 5 columns and 12 rows. Metal inserts make the insertion of electronic parts and connecting wires quite easy.

The advantage of punchboards is that soldering isn't required, mounting and connecting parts is done easily and quickly, and so circuit variations can be tried without too much effort. Kits contain circuit suggestions and possible parts layouts. Independent experimenters need to obtain or develop their own circuit diagrams and parts layouts. Generally, kits make it possible to build a fairly large number of circuits. This type of construction is an excellent way of becoming familiar with the actual appearance of radio parts, with symbols, pictorials, and circuit diagrams.

COMPONENT DENSITY

Component density refers to the total number of parts that can be contained per cubic unit of space. In the early days of radio, parts were large and little consideration was given to the space between them. As a result, radio receivers using a small number of tubes occupied cabinets two feet or more in length. Subsequently some effort was made to miniaturize tubes, resulting in so-called "peanut" and "acorn" types.

Size reduction of parts started with miniaturization, then subminiaturization, and finally microminiaturization. Space became a problem when receivers were to be mounted in aircraft and when a demand began to exist for portable radios. An early effort to improve component density was modular construction consisting of two or more small platforms mounted one above the other on which radio parts were positioned. Known as a module, the vertical arrangement helped reduce the lengths of connecting leads and gave the communications unit more of a boxlike appearance. However, servicing was more difficult than with receivers that used a flat, horizontal chassis, since the parts weren't as accessible.

It wasn't until the development of PC boards and integrated circuits (ICs) that a tremendous breakthrough was made toward optimum component density.

INTEGRATED CIRCUITS

An integrated circuit (IC) also known as a microcircuit, can be regarded as a logical adjunct to the printed circuit. These units may contain thousands of transistors, diodes, resistors, and capacitors, and can be classified as ICs, large scale ICs (LSIs) and very large scale ICs (VLSIs) depending on the number of components. The chief advantage is the very small size of an IC, but if any part of an IC becomes defective, the entire unit must be replaced.

ICs have not made PC boards obsolete. On the contrary, the two are often used together, but the development of the IC has greatly reduced the number of separate components once individually mounted on PC boards. In a certain respect ICs (except for size and method of manufacture) resemble PC boards since the components of an IC are part of a semiconductor substrate that not only supports but is a part of the circuits that are constructed on it.

The IC is based on the fact that solid-state diodes, transistors and resistors as well, can be manufactured from silicon. The conducting paths between these parts is not an etched or painted technique as in the case of a PC board, but is done by a chemical process. Upon completion the unit is referred to as a chip, and when packaged, complete with leads, is known as an integrated circuit, more often called an IC.

CHIPS

A chip is a single substrate (or base) of semiconductor material (sometimes called real estate) on which all the active and passive elements of a circuit have been fabricated. Various techniques are used for this including diffusion, passivation, photo resists, epitaxial growth, and masking. The components on the chip may include diodes, transistors, resistors, inductors, and capacitors. Before the chip can be utilized it must be packaged, and that package must be supplied with terminals (called pins) to which connections can be made. Most, but not necessarily all, of these terminals will be used.

Sometimes the word "chip" is used synonymously with "integrated circuit," but the two are different. The IC is a package, complete with terminals, ready for mounting on a PC board. The chip is the substrate contained inside that package, complete as far as its circuitry is concerned. In some instances, the diodes and transistors fabricated on the substrate are also referred to as chips, but this is at variance with the commonly accepted interpretation of the word chip. The substrate material for the chip is cut from a larger section of a semiconductor called a "wafer." Thousands of chips are made at a time on a wafer, and then the wafer is cut apart to supply individual chips, preparatory to packaging.

While it is technically possible to include inductors and capacitors on chips, it is generally impractical to do so. It is difficult to have capacitors on a chip because of the large area required to produce any significant amount of capacitance, even using material having a large dielectric constant. Similarly, inductors also require a large surface area. For that reason ICs generally consist of transistors, diodes, and resistors, with capacitors and coils used as externally connected discrete components.

Circuit Differences

It is possible to have two indentical circuits, one using ICs, the other discrete components only. If the IC does not contain inductors or capacitors, and this is most likely, then the two circuit diagrams, one using an IC, the other not using it, will appear to be completely different.

However, this is also true of circuit diagrams using discrete components only, but not to such an extent. It is possible, using separate parts throughout, to draw a circuit diagram in a number of different ways. This is sometimes done to emphasize the operation or connection of one or more subcircuits or the relationship of some components to each other.

IC NOMENCLATURE

For use in computers, chips can be categorized as two basic types: processing or storage, depending on the function for which they are intended. Processing chips do arithmetic, or the comparison of numbers, control functions or monitoring. Storage chips, as their name implies, hold information or instructions. They can do this in a format that can be changed, and are identified as RAM (random access memory) or in a permanent format called ROM (read only memory).

For electronics other than computers, ICs can be categorized in a number of ways, but basically they can be grouped under two headings: linear (or analog) and digital. An analog signal is one that is continuous. A sine wave is an example of an analog waveform. The voltage produced at the output of a microphone is an analog of the sound energy at the input to that mic. A digital IC, however, works with waveforms that pulse either on or off. The digit 0 can represent an off condition; digit 1 means on. These two numbers, 0 and 1, are the only two symbols used in digital circuitry, and are the basis of a system of arithmetic called "binary." The much more familiar system, the decimal, has ten symbols ranging from 0 to 9.

ICs for handling binary numbers are used in compact disc players and in computers. Linear ICs are found in tuners, pre- and power amplifiers, and many other electronic components. All ICs, though, whether linear or digital, are mounted on PC boards. Sockets may be used, or else the ICs can be soldered to components or PC board wiring. Digital ICs are also called "logic ICs."

Aside from the fundamental division of ICs into linear and digital, they can also be categorized based on construction. A bipolar IC is one that includes either P-N-P or N-P-N transistors, or both. CMOS ICs are complementary metal oxide semiconductor ICs.

MONOLITHIC ICS

The word monolithic is based on two parts: mono, meaning one, and lithic, meaning stone. A monolithic IC is made on a single (mono) piece of semiconductor substrate (stone). Actually the substrate is not a stone in the true sense of that word and as used here simply means a solid substance.

There are various types of monolithic ICs, including P-N junction isolated, dielectric isolated, and beam-lead. However, unless there is some specific need for a detailed description, these units are simply referred to as monolithic ICs.

ARRAY ICs

One of the advantages of an IC is that it minimizes the amount of point-to-point wiring required in a component such as a receiver or amplifier. The diodes and transistors in the IC are internally interconnected and so there are relatively few connections to be made externally. An IC may have as many as 30 terminals, (or more) but generally the number of connections is much less than this.

The disadvantage of this setup is that it limits the applicability of the IC. If circuit changes are wanted, this often means a search for an IC that will be suitable. An array IC overcomes this difficulty since its substrate carries a number of diodes and transistors, with each of these independent. As a result, each can be regarded as a discrete component (Figure 5.5). A socket is desirable for an array IC for experimental purposes. The IC can be removed at the time the leads are being soldered or unsoldered. Having the transistors or diodes in an IC package is desirable since they are protected. Further, removing the IC package at the time soldering and unsoldering is being done protects the diodes and transistors from heat damage.

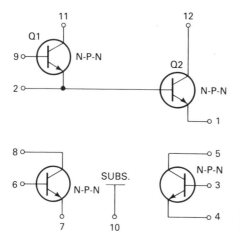

Figure 5.5. Array IC uses independently operating semiconductors.

LARGE SYSTEM ICs

The array IC and an IC called a large system IC are at opposite ends of the IC concept. The array IC contains semiconductors only, either diodes or transistors, but excludes resistors. If the IC has transistors only, it is referred to as a transistor array; if diodes only, a diode array. A large system IC, also called a subsystem IC, packs in as many components as possible. Thus a large system IC might contain all the parts of an entire section of a receiver. This could include demodulator circuitry, an FM limiter, and an audio voltage amplifier. Other large system ICs might consist of an RF and IF amplifier, a mixer, and a local oscillator. A complete receiver, then, could consist of a succession of a relatively few ICs.

PIN STRUCTURE

The pins connected internally to a chip and extending from an IC package may form right angles with that package, but in some ICs the pins extend straight outward, a type of construction known as a flat pack, with the pins soldered directly to conductors or discrete components on a PC board. The advantage of a flat pack IC is that it can be closer to the board if space above the board is at a premium. Unlike PCs that are right-angle pin types, flat pack ICs do not have the option of using sockets. A further disadvantage is that extreme care must be used in soldering and unsoldering. The connections are extremely close, and it is easy to form an unwanted solder bridge between pins, especially since they are pretinned to permit easy soldering.

INDEXING

When a plug has its pins arranged symmetrically, some method must be adopted to prevent mismating with its corresponding jack. This is done by indexing. In some instances indexing automaticaly prevents mismating by some physical means. In the case of ICs, an index is a depression of some kind at one end of the upper face of the IC package. The index may be rectangular or may form an arc (Figure 5.6).

When indexing is done physically, it is referred to as a "polarizing slot." Such a slot, also called an "indexing" slot, may be used by PC boards. One or more of these are placed along the edge of the board so as to align the board with external connectors.

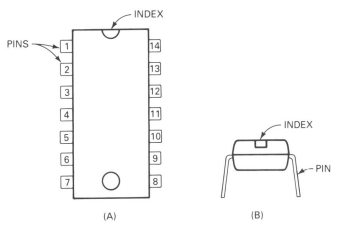

Figure 5.6. Indexed IC. Top view (A) of IC using curved index; side view (B) of IC using rectangular index.

With an IC the index isn't physical, but is simply used to call attention to the order of pin numbering. Holding the IC so that the index is visible, the number 1 pin is at the upper left, with succeeding pin numbers arranged in a counterclockwise manner. Many ICs, however, do not make use of an index.

When used, an IC index will be indicated on the layout diagram of the PC board but is often not included in the circuit diagram. The reason for this is that it is not the function of the circuit diagram to show parts positioning.

The IC Symbols

One IC symbol is an equilateral triangle (Figure 5.7) and appears in circuit diagrams with one apex pointed to the right. The symbol may be numbered, with the numbers indicating connections to be made to external circuitry. Short lines may extend from these connection points but are sometimes omitted.

Figure 5.7.   Symbols for ICs. The triangle is commonly used to represent an op amp. In some instances one symbol is used inside another. Thus, the op amp symbol may be enclosed by an IC symbol.

In some instances the circuitry of an IC accompanies the symbol, but this is usually not done when the IC is part of a component that is represented by a large circuit diagram. The advantage of using the triangle symbol (or the alternate rectangular block symbol mentioned below) is that it simplifies the overall circuit diagram and makes it look less cluttered and less complicated. Generally, the symbol and the IC carry a manufacturer's part number (Figure 5.8). The overall functioning of the IC can be obtained from the manufacturer's spec (specification) sheet.

Figure 5.8.   Noise reduction IC. Dimensions represent those of the enclosure, not the chip itself. (Courtesy dbx, Inc.)

A rectangle is also used as an IC symbol and is positioned in a schematic diagram so that its short sides are vertical; its long sides horizontal. But while it is often used, its positioning isn't standardized. Like the triangle the symbol may carry the manufacturer's part number. It may also be numbered around its sides to indicate connections for external circuitry. Often the triangular symbol is used for op amps (described

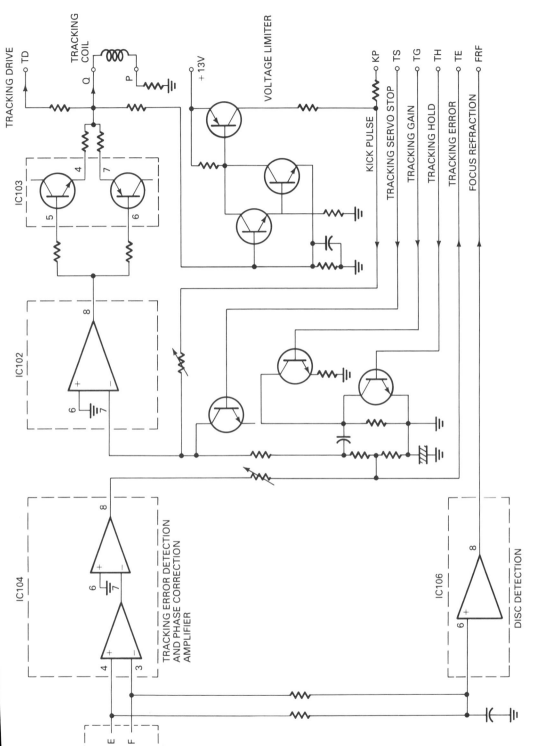

Figure 5.9. Tracking servo block diagram.

later in this chapter) while the rectangle is for other types of circuits; often this symbol distinction is not followed.

In some diagrams, such as Figure 5.9, the rectangular and triangular (or other) symbols are combined into one. The block is shown as a rectangle drawn around the triangular symbols. In this example the triangles represent op amps. If two of the symbols are included within the rectangle, it represents a hybrid IC. This diagram is that of a tracking servo in a compact disc player and consists of a number of ICs and discrete components.

IC Packages

An IC is supplied with the chip securely mounted in an insulated container. The connecting leads are brought out in several different ways. Looking down on the IC from above the output terminals are often numbered in a counterclockwise manner starting with digit 1 in the upper left-hand corner (Figure 5.10A). In some units all the leads are brought out only from one side (B) and the component is then known as a single in-line package (SIP). More commonly, though, both sides of the IC package are used, and the component is then called a dual in-line package, or DIP. Some ICs are in circular form. (C). In some instances all four sides of the package are used.

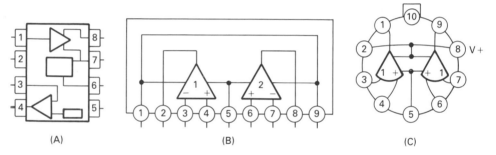

Figure 5.10.   IC packages. Dual in-line (A) package (DIP); single (B) in-line package (SIP); circular package (C).

ICs are identified by a part number printed on the case. However, if an IC is used in a circuit diagram, there is no way of knowing what the IC is or what it is supposed to do. In some instances an overall circuit diagram may be accompanied by a separate diagram revealing the wiring of the IC. These diagrams usually accompany the IC when it is purchased (Figure 5.11). No IC works by itself and must be connected externally to a voltage source and supplied with an input signal. The output can then be taken from one of the terminals on the IC. While the IC may contain just a single op amp, it may also have two or more. In some instances, especially when the IC comprises a number of different circuits, a block diagram is supplied instead. Thus, for an IC we can have four different types of diagrams: the outline diagram of the IC showing the numbered connections; a block diagram using op amplifier symbols only; a block diagram; and finally, an actual circuit diagram, re-

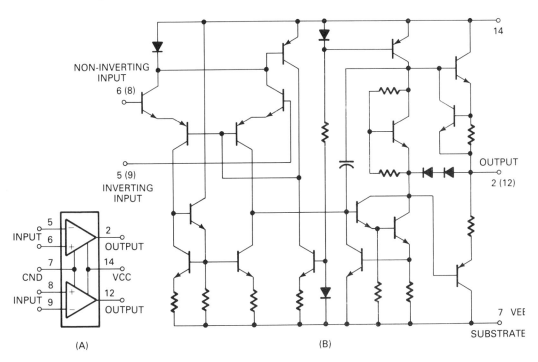

Figure 5.11. IC containing two op amps (A); circuit diagram of the IC (B).

ferred to in this technology as an equivalent circuit. A single block diagram may accompany the IC with a list showing the connections to be made.

The Bump IC

To avoid the need for making soldered connections to the IC, some units are designed with small protrusions known as bumps. These solder bumps provide easy mechanical mounting and electrical connections (Figure 5.12).

BUMP DIAMETER	0.0006″ ± 0.001
BUMP HEIGHT	0.004″

Figure 5.12. Bump IC (Courtesy Cherry Semiconductor Corporation).

Multifunction IC

An IC may contain a number of circuits with these shown in block diagram form (Figure 5.13). This IC is a complete camera system on a single chip supplying shutter timing, a battery check, and a low-light indicator. The circuits include a comparator, an OR gate, an electronic double pole, single throw switch, a Schmitt trigger, and a trio of output driver circuits. The electrical characteristics of an IC such as this one are sometimes supplied as part of the block diagram, printed on a separate sheet.

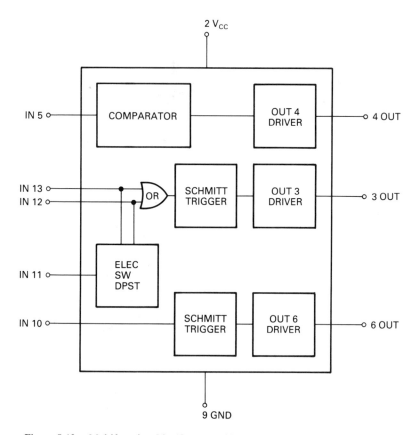

Figure 5.13. Multifunction IC. (Courtesy Cherry Semiconductor Corporation).

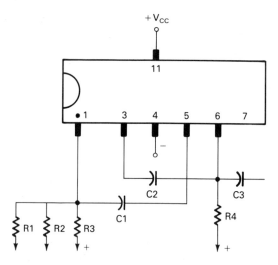

Figure 5.14. External components wired to IC.

In some diagrams the external components are shown wired to the IC (Figure 5.14). These are not supplied in detail, but the diagram is of help when making connections.

DISADVANTAGES OF ICs

There is no question that ICs have many advantages. Used in conjunction with PC boards they permit the easy and quick construction of numerous circuits, and because they consist of prewired parts are an assurance of connection success. A tremendous number of ICs are available for just about every conceivable circuit, and futher, it is often more economical to buy an IC than the electronic parts it contains. From a practical manufacturing standpoint, ICs have kept the prices of radio and TV receivers consistently low.

Considered from the viewpoint of reading circuit diagrams, ICs have made matters more difficult, not less. An IC in a circuit diagram can be a mystery. Often enough, the only information about the IC is a part number printed on it, and since the IC is so small, that may be difficult, if not impossible, to read. In some instances it is nothing more than a numbered symbol, and while its action in a circuit can be deduced, the electronic action that goes on inside the IC is just a guess. Matters are helped if a circuit diagram of the inside of the IC is supplied, but sometimes, especially in large and very large scale ICs, tracing the path of a signal can be tedious and frustrating.

Quite often, too, the circuit of the IC is not supplied. Its behavior may be known, but just as a generality—that is, the IC is an oscillator, or an amplifier, and so on. In some instances the shape of the IC symbol is a clue. Triangles are used for op amps; special symbols are used for ICs containing logic circuits.

HYBRID ICs

In some instances a pair of ICs may be combined in a single package comprising a linear IC plus diodes and resistors, or op amps plus nonamplifier circuitry. These are referred to as hybrid ICs.

IC APPLICATIONS

An IC can be a subcircuit, but more usually is a partial circuit, such as a complete audio preamplifier, demodulator, RF or IF amplifier, oscillator, mixer, and so on. ICs can work in analog or digital components, as op amps (described later in this chapter), as logic circuits, and so on. ICs can include bipolar transistors, FETs of every kind, and different types of diodes. It would be difficult to list, let alone describe, every IC application since ICs are so widely used, but the following few samples of linear ICs should supply a general idea.

IC Voltage Regulator

The regulation of a power supply, in percentage terms, is the amount by which the output voltage will vary with an applied load. Regulation can be improved by including voltage regulator circuitry, and this can be done by using an IC for that purpose (Figure 5.15). The overall circuitry would have been simplified if it had been possible to include filter capacitors C1, C2, and C3, but their large capacitance makes it impossible. C1 is an electrolytic, C2 and C3 are tantalum units, and all of these are polarized. The bleeder resistor R1 must also be an external unit.

The use of the IC does simplify the overall circuit diagram, but aside from the fact that we know this IC is a voltage regulator we know nothing of its circuit arrangement.

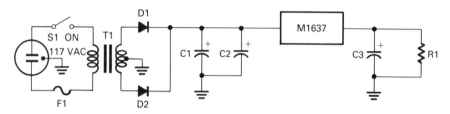

Figure 5.15.   IC used as a voltage regulator in a power supply.

IC Filter

The IC in Figure 5.16 is a filter used in the IF section of a receiver. It is inserted between the output of a dual gate FET mixer stage and the input of dual gate FET used as an IF amplifier.

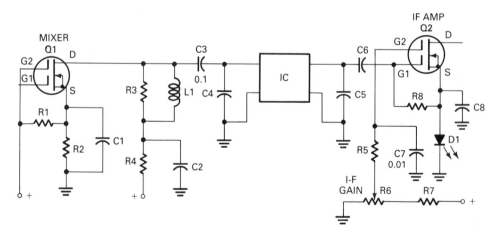

Figure 5.16.   IC used as IF filter.

IC Preamplifier

An IC can be used as a preamplifier (Figure 5.17), working as a voltage amplifier whose output drives a power amplifier. In this example the IC shows the numbering of its terminals, but this practice isn't always followed. The IC part number is supplied directly on the symbol. While the IC manufacturer's name isn't printed on the IC it may appear in an accompanying parts list assuming the parts list is included.

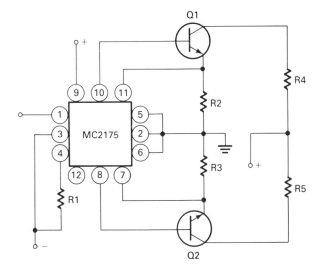

Figure 5.17. IC audio preamplifier.

IC Mixer

In a superheterodyne receiver, the incoming RF signal is heterodyned in a mixer circuit with the signal supplied by a local oscillator. Figure 5.18A shows one type of mixer circuit. This circuit is an IC as indicated by the dashed lines surrounding the unit. Ordinarily the IC isn't portrayed this way, but rather as shown in (B).

SURFACE MOUNT TECHNOLOGY

To understand surface mounted components (SMC), it is first necessary to understand printed circuit boards (PCBs) and their limitations. A PCB is a component, and it is as much of a component as a resistor, a capacitor, a transistor, or any other device used in electronics.

    A PCB can flex, and this can be done mechanically, the extent depending on the material and its thickness. The board can also twist due to temperature changes, the amount depending on the temperature and on the board's temperature coefficient. The temperature coefficent of expansion is the rate of expansion of a material measured in ppm/°C when the material's temperature is increased.

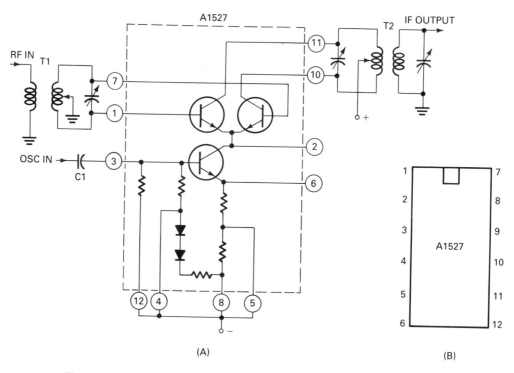

Figure 5.18.   IC used as a mixer. Circuit diagram of IC (A); package arrangement (B).

The printed circuit board, also called a printed wiring board (PWB) is a substrate of epoxy glass, clad material, or other substance, and it is on one side of this base that conductors are formed for connecting all the components.

After the parts are mounted on the PCB, wave soldering is used to make the electrical connections. The board is passed through one or more waves of molten solder, which is continuously moving to maintain fresh solder in contact with the board. The advantage of wave soldering is that all the connections are made at one time, an operation that is more cost effective and faster and produces results that are superior to point-by-point soldering.

A PCB can be single- or double-sided. With a single-sided board all the parts are mounted on one side with their connecting leads passed through previously drilled holes. The reverse side consists of conductors only. With a double-sided board, components are mounted on both sides. The advantage of the single-sided arrangement (the more commonly used) is that connections can all be soldered at the same time using the same soldering technique.

The Need for a New Board Technology

Prior to the introduction of PCBs, components were mounted on a metal chassis above and below this surface, with these connected by individual wires. This method involved considerable hand labor, with wiring errors and poorly soldered connections not uncommon. The PCB guaranteed connection accuracy and reduced manufacturing time.

The PCB was a tremendous advance over the metal chassis but because of advances in electronics technology, the limitations of PCBs have become evident.

What is Surface Mount (SM) Technology?

Surface mount (SM) technology is a radical departure from typical PCB technology. Components and their connectors are mounted on one side of the board only, with the reverse side blank, thus eliminating the need for drilling through holes. The connecting points of the individual parts are soldered directly to the circuit wiring. And while the reverse side is a blank, it remains available for use.

Because of the way the components are mounted and soldered, both the manufacturer of electronic equipment and the servicing technician repairing that equipment require completely new soldering and unsoldering devices. Surface mount (SM) devices cannot be easily removed with a soldering iron or gun, nor can they be replaced using these tools.

The original impetus to the development of SM technology can be attributed to ICs. Manufacturers of in-home electronic portable, mobile, and auto electronic systems must stay competitive by emphasizing operating features. In turn, these put pressure on PC boards to carry more and more ICs. Further, these PCs were packed with more circuitry, requiring additional connecting leads. While IC packages with 8, 14, 16, 18, 22, 24, 28, and 40 pins are not uncommon, some of the newer packages are up to 244 pins.

It isn't the chip itself that has become larger, but the package in which it is contained. The increase in package size is due to the larger number of leads. Further, packages for SM have their leads spaced on 50-mil centers, much smaller than those used on drill-through PC boards.

Consequently, a new PC board technology had to evolve. Without this new technology, electronic entertainment components would have had to become much larger in size, unthinkable for home units and impossible for portables and auto installation. With SM printed-circuit boards can be reduced as much as 50 percent, and at the same time are capable of accepting multipin ICs having well in excess of 40 pins.

Electronic Benefits of SM

Greater component density and a reduction in assembly time are only two of the benefits for manufacturers. SM components use leads having a smaller size, and this means a reduction in lead resistance and in capacitance. For low-frequency use this is probably insignificant but becomes important when high frequencies are involved, or in apparatus where speed of operation is important, as in the case of computers or digital communications equipment.

PCB vs SM Component Mounting

Surface mounted components can include all components that are already familiar: resistors, capacitors, other discrete units, and both single and dual in-line PCs (DIP).

Figure 5.19 shows the difference in the customary assembly of an IC on a PC board and a surface mount assembly. In Figure 5.19A the DIP must be positioned through the pre-drilled holes on the board. Surface mounting completely eliminates hole drilling (B), and the component is soldered directly to the conductors on the board.

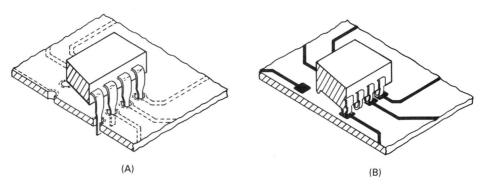

(A)   (B)

Figure 5.19.   Dual in-line package (A) and surface-mounted component (B). (Courtesy Texas Instruments)

Board component density is increased for two reasons. Since the opposite side of the board is unused it can be reserved for a completely different circuit assembly. In effect this doubles the utility of the board and permits more circuits to be included in the same space. Because no space needs to be reserved for through-holes the size of the board can be reduced. In a representative example, a board measuring 10.7 × 14.25 inches with DIP ICs mounted using the standard through-hole technique was able to use a board measuring 9.6 × 6.5 inches with SM components. The area involved was reduced from 152.475 square inches to 62.4 square inches.

Components to be mounted via through holes on a PC board are leaded—that is, they have conducting leads. The leads must then be bent and pushed through the holes on the board prior to soldering on the reverse side. Surface mounted components have shorter, more closely spaced leads, and these are soldered directly to the conductors on the board. Because the through-holes are eliminated, the components to be mounted can be positioned much closer together.

The size of a symbol in a circuit diagram supplies no clue to the physical dimensions of the electronic part it represents. The symbol for an IC, for example, can be larger than the unit itself. For a dual in-line packaged IC, the spacing between its leads measures only 0.1 inch, but even with the IC on hand there is no clue as to the size of the chip it contains; invariably it is much smaller than its package. As the chip has been called on to do more and more, it is not the size of the chip that has grown but rather its container, a problem based on the need to include more leads.

This problem has been met by the development of surface mount technology. One of its techniques is the use of much closer lead spacing of only 0.05 inch compared to 0.1 inch for the DIP. In effect the lead spacing on the DIP has been cut in half. Because of this 50 percent reduction in lead spacing, the size of the package has similarly been cut in half approximately. The amount of size reduction depends on the number of leads extending from the original DIP.

It is possible for ICs to require most of the space available on a PC board when a number of them are used, but the reduction of the ICs' physical dimensions means that either more PC board space has been made available or else smaller boards can be used. Smaller board size is the more likely consequence since this can help reduce the size of the receiver, amplifier or tuner circuitry. This doesn't necessarily mean that the receiver, amplifier, or tuner will be smaller, but it does indicate more room inside. This leaves more space for circuit integration, as for example, a compact disc player combined with a receiver or an amplifier.

SM Variations

About as fast as a new part development takes place, a number of variations are also made available. And so we have three different types of surface mount devices: the Small Outline Integrated Circuit (SOIC); the plastic lead chip carrier (PLCC); and the Leadless Ceramic Chip Carrier (LCCC) (Figure 5.20).

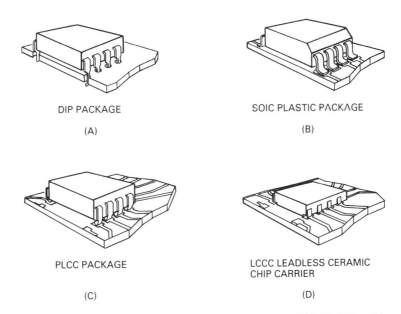

DIP PACKAGE

(A)

SOIC PLASTIC PACKAGE

(B)

PLCC PACKAGE

(C)

LCCC LEADLESS CERAMIC CHIP CARRIER

(D)

Figure 5.20. Package types. DIP (A); SOIC (B); PLCC (C); LCCC (D). "Copyright 1986 Gernsback Publications. Reprinted with permission from May 1986 Radio-Electronics."

Of these three, the SOIC is possibly the most widely used and is characterized by gull wing leads with these spaced 50 mils apart. The advantage of the gull wing structure is that the IC can be positioned flat against the PC board. With this arrangement the package is such that access is easy for testing and repair. Typically, the SOIC package has 8, 14, 20, 24, and 28 pins with these arranged in DIP fashion; that is, the leads are positioned in a dual in-line manner.

One difference between the SOIC package and the PLCC is the way in which the leads are shaped. Instead of using a gull wing design, the PLCC has its leads in J shape, with the J curve tucked under the package. Because the leads are arranged in this manner, the area occupied by the package is reduced, permitting greater component density. This setup limits access to the pins and so troubleshooting is more of a problem. However, unlike the SOIC design, the PLCC package has leads on all four sides. The number of pins used by a PLCC is typically 18, 20, 28, 44, 52, 68, and 84 pins.

Unlike SOICs and PLCCs, the LCCC has no leads. These are like the bump ICs discussed earlier and have metallic contacts positioned on the bottom surface of the package. This doesn't mean the LCCC isn't soldered, because it is. The soldered connections, however, aren't visible because they are covered by the body of the package.

Discrete SM Components

While emphasis has been placed on ICs, both active and passive discrete components can also be used in SM technology. Figure 5.21 shows the use of a single transistor housed by itself in an SOIC package. The arrangement is no different than that of an IC except that fewer connecting pins are involved. The package uses the typical gull wing leads of SOICs and is surface mounted. Unfortunately, the size of the drawing can supply a wrong conception of the size of this unit. Its largest side is only 0.25 inch. The package is an epoxy body and the transistor itself occupies just a small fraction of the interior space. The unit has only three leads: one each for the emitter, base, and collector. It may seem odd to have a single transistor, but not only does every circuit require them but passive components such as resistors and capacitors,

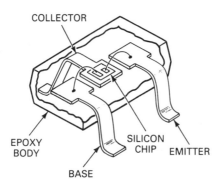

Figure 5.21. Single transistor in SM SOIC package.

and some small coils as well. These resistors and capacitors, often called chip resistors and chip capacitors, are available as surface-mounted packages.

Figure 5.22 shows a typical chip resistor outline—(A) is a top view, (B) is a side view, with the dimensions indicated in inches. The ends of the unit are metallized and are pretinned to facilitate soldering. The resistance element (C) is a glass passivated thick-film element on a high purity alumina substrate. The result is a highly reliable, precision component. Typical resistance ranges are from 10 ohms to 2.2 megohms with either a 1 or 5 percent tolerance. Power dissipation is rated at 1/8 watt.

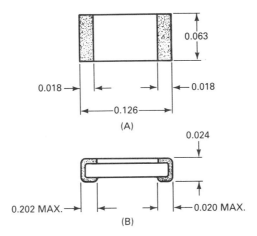

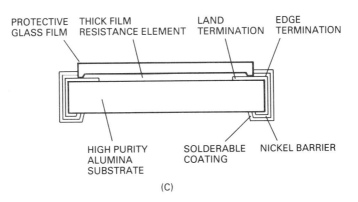

Figure 5.22. Chip resistor for SM use. Top view (A); side view (B). All dimensions are in inches. Cross-sectional view (C). (Courtesy Texas Instruments)

Figure 5.23 illustates a chip capacitor with a totally encapsulated electrode system, and, like the chip resistor, using metallized terminations. The electrodes, the metal plates that form the capacitor, are interleaved. The layers are stacked to increase the capacitance to the required value.

TERMINATION

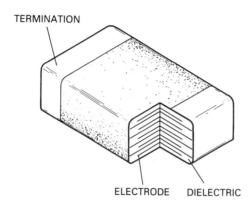

ELECTRODE   DIELECTRIC

Figure 5.23. Chip capacitor for SM use. (Courtesy Texas Instruments)

Note that neither the chip resistor or capacitor have leads extending from the component, a type of construction used with typical PC boards. For surface mounting extension, leads are not required. Capacitances as high as 0.47 μF at 50 volts and 0.33 μF at 100 volts are obtainable. These values depend on the type of dielectric used. Other capacitances such as values up to 4,700 pF at 100 volts and up to 10,000 pF at 50 volts are also available.

The physical dimensions (length $\times$ width $\times$ thickness) for chip capacitors range from 0.080 $\times$ 0.050 $\times$ 0.050 to 0.175 $\times$ 0.125 $\times$ 0.060 inch, but several intermediate package sizes can be had.

Hybrids

The transition from through-hole type PC boards to boards designed for surface mount components isn't one that is sharp. The reason for this is that there are numerous eletronic parts not yet designed for surface mounting. On the other hand, resistors, capacitors, field-effect and bipolar transistors and diodes are available in surface-mounted packages. But each of these is just one of a long list. For each model number there can be dozens of other model numbers.

When a certain component is not available as an SM device, a manufacturer may have no option but to use what SM parts he can obtain with all the others to be mounted with their leads pushed through holes on the board. As more components are made, and as surface mounting becomes more prevalent (as it will be) these hybrids will disappear. Ultimately we will have PC boards that are completely SM for all components. A broad line of ICs in SOIC, PLCC, and LCCC packages are available to support the development and production of surface mount assemblies. A few of these include bipolar and CMOS logic and memory products, advanced low-power Schottky and advanced Schottky devices.

Inspection

It might seem that inspection of the soldered connections following the manufacture of PCBs carrying SM devices would be difficult. This is true of visual inspection since the leads are covered by the package in the case of PLCC and LCCC. And even with SOIC packages, the solder connections are hidden. However, the inspection of soldered connections can be done using x-rays and laser beams.

Testing Surface Mount ICs

Since ICs are extremely small, testing them when they are mounted on a PC board can be difficult. The pins are tiny, so positioning a test probe, even a needle-point type, is extemely difficult. Further, with extremely fine lead spacing it is easy for the probe to slip and short out adjacent leads with possible damage to the SM device. The connecting leads in the case of PLCCs and LCCCs positioned beneath the IC being tested make servicing difficult. Further, to obtain maximum component density, some ICs butt-joint each other leaving no access room.

One method of reaching in to the SM units is to use an IC test clip. This makes a close fit over the IC under examination and has more widely spaced pins protruding from its upper surface. However, a single IC test clip will not be sufficient since ICs can have different packages. For the service technician it will be necessary to have a variety of test clips of different sizes and shapes.

Footprints

The term "footprint" is used in satellite TV technology to indicate the geographical area coverage of a satellite signal. In SM component mounting, it refers to a group of tiny metal areas or fillets on the PC board arranged in an identical pattern as the leads of an SM device. Ultimately these leads will be soldered to the corresponding footprints on the board (Figure 5.24).

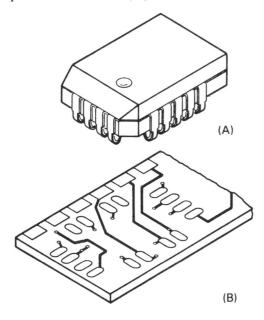

(A)

(B)

Figure 5.24. Top view of surface-mount component (A). Footprints on the PC board (B). During manufacturing the leads of the SM will be made to mate with the footprints. (Courtesy Texas Instruments)

The first illustration (A) is that of a PLCC package; (B) shows the fillets. These look like ellipses, with each one representing a tiny soldering area. Note that not all the fillets are connected to conductors. The particular IC in this example has 16 fillets, but not all of them are used.

Static Charge Removal

The buildup of an electrostatic charge on an individual's hand can be in the order of several kilovolts or more, easily capable of damaging board components. To dissipate this charge, it is necessary for both the work surface and the technician to be at ground zero potential. This can be done by using a conductive work surface that is grounded, and a wrist strap or band that is connected to it. The static discharge element can be an expandable wrist band (Figure 5.25). It also works to protect the wearer from eletrical shock. While the work surface (Figure 5.26) remains permanently grounded, the static watch can be easily removed completely or detached from its connecting wire. The watch band can be mounted on either wrist to accommodate those who are left- or right-handed.

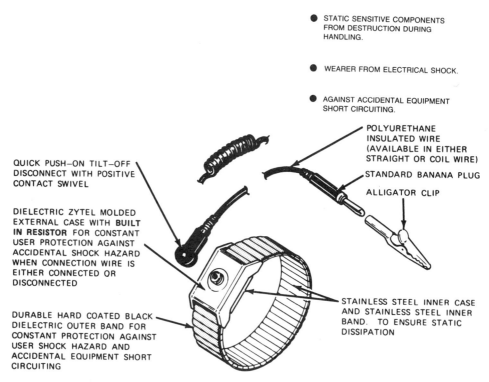

● STATIC SENSITIVE COMPONENTS FROM DESTRUCTION DURING HANDLING.

● WEARER FROM ELECTRICAL SHOCK.

● AGAINST ACCIDENTAL EQUIPMENT SHORT CIRCUITING.

POLYURETHANE INSULATED WIRE (AVAILABLE IN EITHER STRAIGHT OR COIL WIRE)

STANDARD BANANA PLUG

ALLIGATOR CLIP

QUICK PUSH-ON TILT-OFF DISCONNECT WITH POSITIVE CONTACT SWIVEL

DIELECTRIC ZYTEL MOLDED EXTERNAL CASE WITH BUILT IN RESISTOR FOR CONSTANT USER PROTECTION AGAINST ACCIDENTAL SHOCK HAZARD WHEN CONNECTION WIRE IS EITHER CONNECTED OR DISCONNECTED

DURABLE HARD COATED BLACK DIELECTRIC OUTER BAND FOR CONSTANT PROTECTION AGAINST USER SHOCK HAZARD AND ACCIDENTAL EQUIPMENT SHORT CIRCUITING

STAINLESS STEEL INNER CASE AND STAINLESS STEEL INNER BAND. TO ENSURE STATIC DISSIPATION

Figure 5.25. Static discharge expandable wrist band. (Courtesy Nu-Concept® Systems, Inc.)

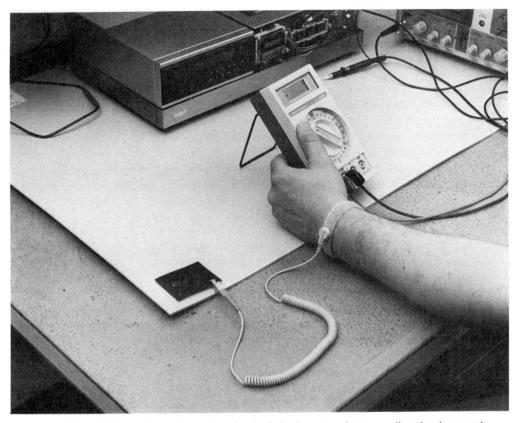

Figure 5.26. Antistatic kit consists of static dissipative mat, wrist strap, coil cord and a grounding cable. (Courtesy RCA)

Conductive and Convective Heating

There are two methods for removing and replacing SM components—conductive or convective heating. A conductive technique is one in which a high temperature is directly applied by a heat source. A common example is the heated tip of a soldering iron or gun which makes direct physical contact with a soldered surface. Convective heating is a hot air method. There is no direct contact, but a flow of heated air is used for solder reflow. Both methods, conductive and convective, are used for the removal and replacement of SM components.

Conductive Methods

Conductive heating using a soldering iron is a point-by-point method. This means that only one connection at a time can be worked on. Further, depending on the way the leads of the package are connected to the board wiring, it may not even be possible for the iron to reach the soldered terminals. Even if the terminals are accessible for a multiterminal device, this would not only be time-consuming but could result in destruction of some of the wiring on the board and possible damage to the board itself or adjacent components.

Instead of a soldering iron, use a heat probe, a conductive tool that can be used to remove or replace any chip package: SOIC, PLCC, or LCCC. It is capable of heating all the terminals simultaneously, and since there are a variety of chip package forms, it comes equipped with a large number of differently shaped heating elements with each capable of being mounted quickly on the probe. Thus, one shape might be square, another rectangular, and so on. Further, these geometric arrangements could all be of different sizes.

A representative probe (Figure 5.27) using conductive heating could have a pair of L-shaped brackets with these forming a square or rectangle. These are positioned at the end of the probe, which has a scissor-like action. The jaws of the probe are opened, and the brackets are made to surround the terminals, making good physical contact. Since the components that are surface mounted are quite small, so are the heating brackets. A pair of square-shaped brackets, for example, could have dimensions of 1.5 inches, but some are just 0.185 inch.

Figure 5.27. Conductive heater uses pair of L-shaped brackets that fit on all four sides of IC. (Courtesy Nu-Concept® Systems, Inc.)

The heat probe can be plugged directly into the AC power line or else it can be used in conjunction with a control station. The amount of heat to reach the brackets can be adjusted by a knob on the front of the unit. The control station can also act as a holder for the heat probe.

To use the probe, allow the two L-shape tips to preheat. Then grasp the component to be removed with a very light touch, allowing the tips to rest directly on the solder joints. Grip the component gently to avoid damaging it. By preheating and then turning on the heat probe, the solder melt is rapid, reducing the chances of thermal shock to both the board and the component. Properly done, heating will be confined only to the component being removed and will not weaken the bonds of surrounding parts.

SM components can be held to the PC board in two ways: by the solder connections alone or by these connections plus an adhesive. This adhesive is crystalline and giving it a slight twist along its horizontal axis will cause the adhesive to fracture, permitting removal of the component following desoldering. Note that this technique results in unsoldering all points at the same time.

It will not be necessary to rotate the probe tips since these nickel L-shaped brackets have their own heating elements and are designed to contact and melt the solder on each side of the component being removed. The tubes of the heating element are mounted on handles having a bidirectional hinge mechanism. This allows the tips to align with the sides of the chip carrier (package) connections by pivoting slightly laterally.

Prior Treatment

Physical objects are affected differently by temperature; some expand more, others less; this is described as their coefficient of expansion. The LCCC package is ceramic and the substrate of the board is glass epoxy, but these two different materials do not expand equally. It would seem offhand that since the IC package requires so little surface area that expansion due to temperature would be insignificant, but it is enough to cause the development of a shearing force, resulting in a shifting of the metallic contacts of an SM device from the footprint. The problem of thermal expansion isn't applicable to SOIC and PLCC packages since their metallic leads absorb any physical shift. The cure is to make packages and PC board substrates out of the same materials or else to use materials whose temperature coefficient of expansion are so similar that the susceptibility to temperature changes is insignificant. LCCCs are the least popular packages for SM use.

A PC board can be heated with a hand-held hot-air blower. Do this also with boards that have power transistors and so make use of heat sinks, or if possible and convenient, remove the sinks. These counteract the heat from the probe and may make it necessary to hold the heat probe in place for a longer time with possible damage to the conductors or to adjacent components.

It is essential that optimum heat transfer be made between the probe tips and the soldered connections of the component. As a prior step before beginning work with the heat probe, apply a liquid flux to all the solder junction points on all sides of the component. The flux is available in a plastic squeeze bottle having a fine, extended tip. This permits the application of the flux in hard-to-reach areas.

The flux has a double purpose. As in single point soldering, the function of the flux is to prevent oxidation of the metallic surface being soldered. Oxidation not only hinders the soldering process but can prevent it completely. The flux also aids in heat transfer, making it more efficient from the probe tip to the soldered point.

Temperature Indicating Liquid

Inevitably, there are two questions when a heat probe is used. The first of these is how hot the probe should be allowed to become. And the next is a question of when to remove the SM component.

There are different ways of knowing how much heat to apply when using a convection heating method. The desoldering machine, as in Figure 5.27, may have an adjustable control and the setting of this control can be made through experimentation and practice. The manufacturer of the unit may also make a number of suggestions as to how the control should be set.

Another technique, very simple and effective, is to apply a small amount of a heat sensitive liquid to the top of the SM component. This liquid, normally milky colored, becomes clear when the temperature of the SM device reaches within 1° of the required heat.

With experience a technician will be able to get the feel of the heat probe and will be able to determine just when to lift up on the probe and pull the SM component out of its position.

The maximum time you should keep the heat probe (sometimes called a collet) turned on should be about 10 seconds, preferably less. After you remove the probe you may note some excess flux on the board. Remove it with some denatured alcohol. It is true that you may be more comfortable with removing an SOIC surface-mounted device since you will be able to see the gull wings. However, the much more common PLCC can also be removed just as easily, but it may take a bit of practice.

Metal Shielding

If you are concerned about the possibility of heat damage to other components, shield them with sections of aluminum. If these are in the form of boxes they can be easily dropped over any area consisting of one or more SM devices.

Reflow

The SM component is held in position with its leads soldered to its corresponding footprints on the PC board. This solder, of course, is in solid form. The process of heating the solder so that it becomes liquid is known as solder reflow, or simply reflow. Reflow takes place under two conditions: when the SM device is being removed and when it is being replaced. Reflow must always take place simultaneously at all leads.

Soldering Rules

The rule in soldering component leads has always been (a) that the component must first be made mechanically secure and (b) the only purpose of soldering is to make a good electrical connection. While these goals are desirable when using a soldering iron and a through-hole PCB, it has resulted in a problem for the technician. When replacing a single-point soldered lead, for example, it is first necessary to soften the solder to the point where the mechanical connection must be undone. Following this the solder must be removed either through the use of a wick or a solder sucker. Unfortunately this technique sometimes results in damage to that part of the board adjacent to the component being worked on.

With SM components the technique is quite different. All that is necessary is to reflow all the solder points. If an IC, for example, has 24 connections, all these must be solder reflowed at the same time, but with the proper tool this can be done in a matter of seconds. Removal of an SM component with 40 terminals is easier and faster than soldering iron removal of a PC board resistor using just two leads in a through-hole arrangement.

Two-Sided SM Components

Removing a part that has leads on just two sides (Figure 5.28), and this includes dual in-line packages, chip resistors, and capacitors, requires the same technique as that used for components with terminals on four sides. To remove these parts it will be necessary to change the probe tip. Select the correct size tip, something you can do by comparing the tip with the length of the SM component.

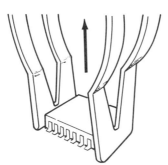

Figure 5.28. Conductive heating collet for multiterminal two-sided package removal. (Courtesy Pace, Inc.)

Solder Fillets

A solder fillet is the concave junction formed by the solder between the footprint and the SM component lead. Sometimes the amount of solder remaining is very small or may be completely missing, and so it will be necessary to add more solder to what was originally the fillet. Additional solder is necessary (following the treatment with flux) not only for making an electrical connection but also to conduct heat to the winged lead of the SM component. This lead will not accept solder unless its temperature is about the same as the melting point of the solder. In some instances the lead of the SM device is narrower than the footprint. In that case, the solder added to the fillet helps spread heat uniformly around the edges of the lead. The rule in remounting a replaced SM component is "get in and out as fast as you can." To be able to do so you must rely not only on the heat probe but you must also make working conditions suitable for the quick spread of heat from the probe.

Convection Heating

The use of a conductive heat probe may not always be possible. In some instances not only the ICs, but other components as well, may be positioned extremely close to each other in an effort to achieve maximum component density. The component mounting technique depends entirely on the manufacturer.

Convection heating is an alternative method for removing and replacing SM components. It consists of a machine (Figure 5.29) that uses a stream of hot air with that air flow reaching all the connections. Unlike the heated probe there is no physical contact between the device producing the heated air and the SM components.

The hot air, under low pressure, causes the solder on the connections to reflow, and the SM unit can then be removed from the PCB using a pair of tweezers or any other tool that can hold the SM device securely. The lifting action should be done as soon as the solder reflows, with the tool holding the unit first moved slightly sideways so as to break the surface tension of the solder holding the SM component to the PC board.

The hot air tool can remove and reattach components of all sizes and types. It can also be used to preheat the board and uses low-pressure hot air (so as not to move small components). The machine has an air blower with two controlled heat tubes that heat both sides of the PC board simultaneously. A platform is provided

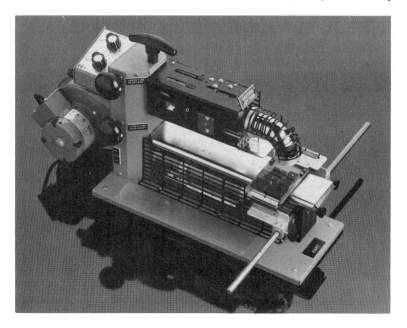

Figure 5.29.   Hot air repair terminal for removing and reattaching components to circuit boards. Temperature indicators assure circuit board and components remain under 400 °F during rework. Device has platform or holding PC board while work is being done. (Courtesy Nu-Concept® Systems, Inc.)

between the tubes for the circuit board to rest upon and is equipped with a circuit board holding device to clamp the board in position, hence allow hands-free operation.

Prior to use, apply a temperature-indicating liquid to the top of the chip carrier. Initially this liquid will solidify but will turn to a clear liquid within 1 percent of a given temperature. It is available in bottle form for 363°, 388° and 400°F. When the temperature indicator turns to a clear liquid, the solder fillets will be molten and the chip carrier (package) can then be removed.

The hot air tool (Figure 5.30) is best suited for components that are on boards that are very low in density and do not dissipate heat rapidly. To remove a surface mount device, direct the flow of hot air at the solder joints, which hold the component in place. Take care to avoid weakening the solder joints of surrounding components. Use a circular motion to attain a solder melt and to break any adhesive bond that may hold the component in place. Figure 5.31 shows the technique to use to remove a chip resistor or capacitor.

Although the heat from this hot air tool is sufficient to melt solder, it is not a recommended method for reflow soldering when using solder paste. Solder paste may be used (Figure 5.32) when remounting a component to supply a new solder base and also for holding the component in place. Air flow from the hot air tool may be too great and may cause the solder paste to be blown, possibly resulting in a solder bridge. Also, if the heating rate is not even at both ends of the component, tombstoning may result. This is an effect that occurs when one solder joint reflows

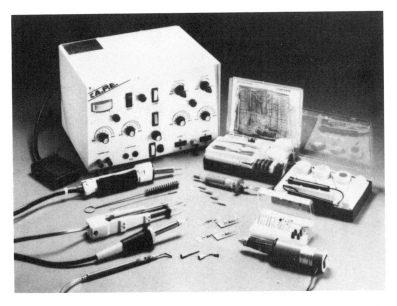

Figure 5.30.   Complete hot air SM device and PC board rework center. The unit is equipped with a vacuum handling tool for the removal or precise replacement of components. (Courtesy Automated Production Equipment Corporation).

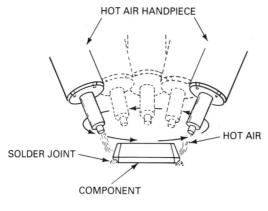

HOT AIR HANDPIECE

HOT AIR

SOLDER JOINT

COMPONENT

Figure 5.31. Use circular motion with this hot air system to reflow solder. (Courtesy Automated Production Equipment Corporation)

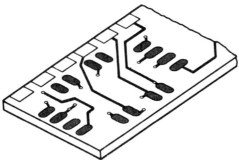

Figure 5.32. Solder paste applied to footprints on PC board. (Courtesy Texas Instruments)

and the surface tension causes the component to stand on end before the second solder joint reflows.

The complete air flow system in Figure 5.29 is equipped with a vacuum handling tool for the precise control and placement of surface mounts.

Still another technique is to use a suitable length of nonkinking wire and insert it between the leads of the SM component and the body of the package. The component is then heated with hot air and when the solder reflows the wire is pulled. This action will separate the package from the PC board.

Preparatory Replacement Technique

Servicing a surface mount device involves three steps: (1) removal by either a conductive or convective method, (2) examination and (3) repair of the footprint on the PC board and of the replacement of the component that has been removed.

The usual rule is that no SM component should be removed unless it is inoperative. But the fact that it does not work doesn't necessarily mean that it is defective. It may not have been mounted correctly, one or more of the solder connections on its pad may make no contact with its corresponding solder point on the footprint, the solder point may have become corroded. There is also the concept, a myth, that removal of an SM device ruins it and makes it useless for replacement.

Independent and user studies have been made to evaluate the effectiveness and integrity of boards and components after the removal and reapplication process. One user removed a component from 50 multilayer boards and resoldered the same component in place. The boards were then tested in the same manner as production run boards, and all boards successfully passed the tests. Another user removed the same board from a computer 20 different times and removed and replaced the same component. Each time the board was replaced in the computer and normal operation of the computer was obtained. The same type of test was performed by another user who removed and replaced a gate array five times and the result experienced was the same.

Before discarding an SM device as defective, then, it is essential to examine the pad and footprint to see if there is some fault that would keep the device from functioning.

Component Mounting by the Manufacturer

While SM component mounting begins as a manufacturing function, the techniques used and the spacing of the components can affect the way in which the service technician will remove and replace them. One production method is the use of a solder paste. The SM devices are then positioned in place automatically by machinery. The paste holds the package in place, and the assembly is then heated to 210 °C, enough to melt the solder. Upon cooling, the solder not only forms the electrical connection but also holds the device in place.

Heating of the solder paste is done by a convection method with the board and its mounted components passing through a heated vapor. The vapor is an inert gas that does not affect the board or its mounted components. It is the surface tension of the solder paste that holds the components in place until soldering is completed.

Whether the manufacturer has the service technician in mind during the design of an SM board is a debatable point.

Servicing Following SM Component Removal

After removing the SM component, inspect all the solder points on the board (fillets), collectively referred to as a footprint. Sometimes removal of the component takes an excessive amount of solder with it. As a result when the part is replaced there will not be enough solder to hold it in place. Further, some fillets will not be connected or will be poorly joined, resulting in inadequate or intermittent operation. Examine each of the fillets and spot solder as required, but do so sparingly. Excessive solder can result in solder bridges from one point to the next.

After you are satisfied that the solder points on the pad on the component are good, and that the corresponding points on the footprint are what they should be, match the component against the footprint. Some SM components can sometimes be mounted incorrectly; others have an index to help prevent this from happening.

Put some liquid soldering flux on each solder point. Use only solder flux made for this purpose since ordinary solder containing a rosin core will not do.

The flux should hold the SM device in place, but sometimes there will be movement of the unit during the heating process. To avoid this put a quick-setting adhesive on the bottom of the SM component. Once the adhesive has set, put the collet frame of the heat probe around the SM and apply heat. You will know the solder flux has vaporized since there will be a tiny puff of smoke. This indicates the solder has liquefied and has reflowed. Follow the same procedure if you are using hot air for solder reflow.

It is important to have the pad of the replacement component adjusted so that its solder points correspond exactly with those of the footprint. This isn't difficult with an IC having about a dozen solder points, but a little more tricky when 40 or more of those points are involved. It does take some practice and experience.

The flux that is used is a granulated solder in the form of a thick paste. With the application of heat the flux is vaporized, with the granules of the paste joining with the solder points of the footprint and the pad.

After the new SM part is soldered into position, do not move it immediately. Wait about a minute for the solder to cool and solidify. Rushing to check by touching the unit may disturb one or more solder points, resulting in cold solder joints.

In some instances poor or intermittent operation may be due to no fault of the component but to one or more of the soldered connections between it and the footprint. A quick way to check is to apply enough heat to resolder. This may solve the servicing problem without the need for removal or replacement.

It is difficult to state which SM soldering method is preferable. Both are desirable. Convection or conduction is a matter of personal preference. It is advisable, however, to avoid any method that requires working by unsoldering connections one at a time.

OP AMPS

An op amp (an abbreviation for operational amplifier) is a high-gain, direct-coupled differential amplifier constructed as an integrated circuit, although it can also be made from individual components. Various transistors can be used, notably bipolar, MOSFET, and JFET, with just one type selected or an intermix. The op amp has a number of terminals, but its symbol frequently indicates just 3 to 14. When the terminals are identified, they usually include the input, output, and applied voltages. Commonly the apex of the triangle symbol for an op amp ordinarily points to the right, but other symbol positions are also used.

Op Amp Symbols

Figure 5.33A shows the basic symbol for an op amp. It consists of a triangle, and usually the plus and minus terminals are indicated. In some circuits a number of op amps may be used. In (B) the number 1/4 inside the triangle indicates there are four such amps in the circuit, with this symbol representing just one of them. Although all four symbols would appear in different places in the circuit diagram, the ICs they

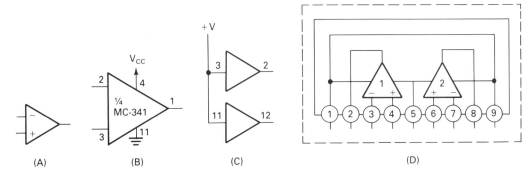

Figure 5.33. Basic op amp symbol (A); partial op amp (B); multiple op amps (C); op amps contained within an IC symbol (D).

represent would be housed in a single package.

In some instances, though, when two op amps in a single package are used, they can be represented by a pair of triangles placed adjacent to each other (C). Since the symbols are so close it isn't necessary to indicate their total by writing a number.

Still another way of showing a pair of op amps is the method in (D). The dashed outline represents an IC package, and the use of the triangles indicates a pair of op amp ICs. Thus we have a symbol inside a symbol.

CHARACTERISTICS OF THE OP AMP

Op amps have a high input impedance and as a result impose a very light load on the circuit delivering a signal. The output impedance is low and the gain is high. Op amps also have a wide bandwidth and excellent operational stability.

While the op amp symbol is a simple one, the op amp itself can be quite complex. Since a complete circuit diagram might include a number of op amps, it could become very complicated if each op amp circuit appeared in complete form. Since the op amp functions as a unit, this is one of the instances in which there may be little reason to be concerned with all the details of its circuitry.

OP AMP CIRCUITS

The op amp is a special form of IC, and while integrated circuits have been supplied in the preceding pages of this chapter, it is helpful to study some of their uses. Like other ICs, op amps require a source of operating voltage and are externally connected to discrete components.

Monophonic Power Amplifier

An op amp can be used as a power amplifier (Figure 5.34). This is a single-ended stage intended for monophonic reproduction only. The IC that is used is an eight-

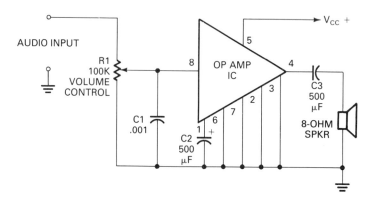

Figure 5.34. Op amp IC used as monophonic audio power amplifier.

pin DIP with four of its connections directly grounded. Its power handling capability is limited to one watt.

Capacitor C1 at the input to this amplifier is in shunt with volume control R1. As a result this combination also works as a tone control. Its effect is to bypass higher audio frequencies making bass tones seem emphasized by comparison. The output of the IC is capacitor coupled, via C3, to the voice coil of the speaker. This IC can operate within a voltage range of 3 to 18 volts, with 9 volts as a typical DC supply unit.

Op Amp Mixer

In Figure 5.35, a linear IC is used as a mixer. The RF signal is applied via a transformer having a tuned primary and a tuned secondary to the input terminals, 1 and 5. A signal from a local oscillator is fed into terminal 2. These two signals are mixed in the op amp and brought out via terminals 6 and 8. Four signals are present at the output: the RF and local oscillator signals and the sum and difference frequencies of these two. The selected intermediate frequency (IF) is tuned by T2 and then brought into following IF amplifier stages.

Op Amp IF Amplifier

A receiver may have a number of IF amplifier stages all using op amplifiers. These amplifiers may be contained in a single package, and so the entire IF stage could be represented by a rectangle containing a number of triangles.

The drawing in Figure 5.36 is that of a single IF stage. The input could be from a preceding mixer stage. T1 is the output transformer, and in this example has just a single polyiron tuning slug for the primary winding. The secondary is untuned.

R1 and C3 form a decoupling filter and the same is true of R2 and C5. Since the op amp uses a single power supply, decoupling filters are used to keep the signal output from leaking back to the input via the common impedance of the battery or DC power supply. Signal feedback could result in amplifier instability, narrowing the amplifier's bandpass, and in extreme cases producing whistling.

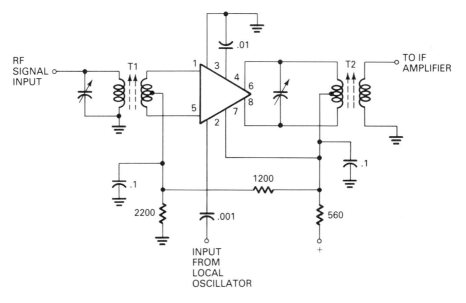

Figure 5.35. Op amp mixer circuit.

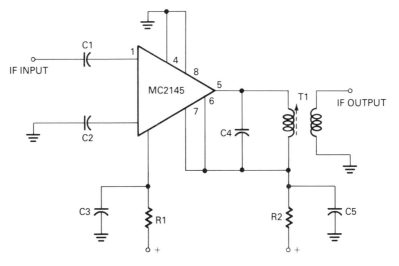

Figure 5.36. IF amplifier using op amp.

6

Partial and Complete Block and Circuit Diagrams

When analyzing a fundamental, sub- or partial circuit it is often helpful to know its general relationship in the overall picture, something that can be supplied by a block diagram.

THE AM RECEIVER BLOCK DIAGRAM

The basic circuit for receivers is the superheterodyne used in all AM, FM, and TV receivers. For AM (Figure 6.1) it includes an RF amplifier, a mixer, local oscillator, IF stages, a demodulator (detector), and audio circuitry. There are variations—the RF amplifier can be omitted and the mixer/local oscillator can be replaced by a converter. The audio output can be single-ended or push-pull. The speaker system can be a single unit for mono or a pair for stereo. This diagram is horizontal (except for the local oscillator), while add-ons, such as headphones, may form a small vertical section.

THE FM RECEIVER BLOCK DIAGRAM

This block diagram is more elaborate than that for AM since more add ons are possible (Figure 6.2). The basic FM unit is positioned horizontally; the add-ons vertically.

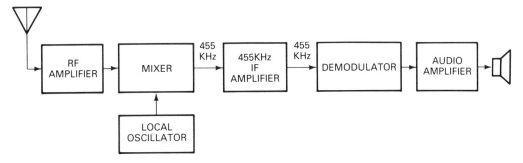

Figure 6.1. Block diagram of AM receiver.

The diagram is simple, and in this illustration consists of the antenna, the AM/FM/MPX tuner including the IF and demodulator, integrated amplifier, and speaker systems. All other blocks are add-ons that may or may not be included. This block diagram can be more detailed because it may have additional speaker systems or other units.

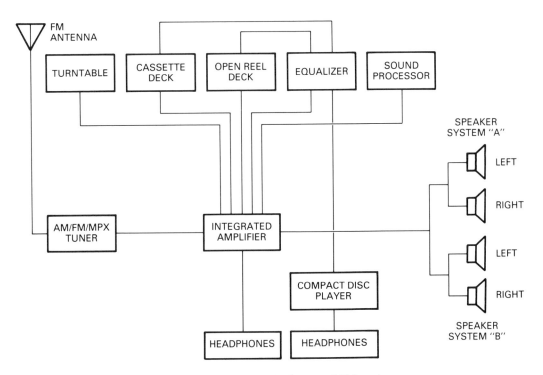

Figure 6.2. Block diagram of FM receiver.

THE AUTO RADIO BLOCK DIAGRAM

Like an in-home FM receiver, an auto radio can be integrated or can be made up of a number of individual components. The auto sound system can be nothing more than a receiver and a pair of speakers. Quite often the receiver is fed into a separate power amplifier (called a main amplifier) for further amplification of the audio signal (Figure 6.3A). The system can be made a little more sophisticated by adding an under-dash tape player, or, quite commonly, the tape player can be part of the receiver (B). A still more elaborate block diagram (C) includes the use of front and rear speakers.

Neither this block diagram, nor any of the others, discloses any information about the installation of the components. For an auto radio, available areas include

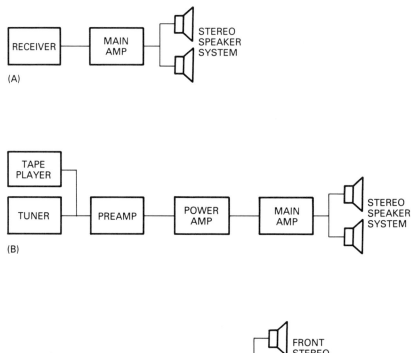

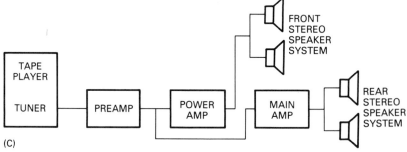

Figure 6.3. Block diagrams of various auto sound systems.

space behind the dash, beneath it, in the console, with speakers mounted in the car's doors and beneath the rear deck. Trunk space is also used for power amplifiers.

SATELLITE TV SYSTEM BLOCK DIAGRAM

Satellite TV signals begin with a dish, usually a parabolic structure pointed directly at a selected satellite some 22,300 miles above the equator. The dish, functioning solely as a reflector and sometimes mistakenly called an antenna, reflects the signal to a feedhorn, a short length of tubing called a waveguide at the far end of which the actual antenna or probe is positioned (Figure 6.4). The antenna supplies the signal to an amplifier called a low-noise amplifier (LNA) followed by a downconverter, a circuit used to lower the signal's carrier frequency. This signal is delivered via coaxial cable to an in-home located satellite receiver, which supplies the signal to a TV receiver.

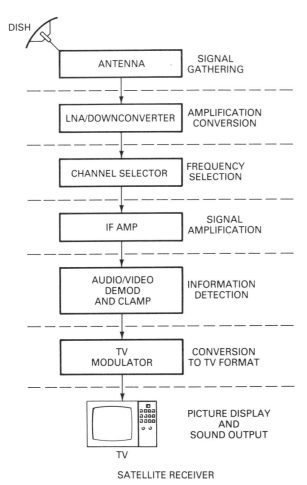

Figure 6.4. Satellite television receive only (TVRO) system.

BLOCK DIAGRAM CLARITY

A block diagram is an explanation of circuit functioning, and, like a verbal exchange, can be readily understandable or confusing. For that reason the best block diagram is one that has been planned to supply the maximum information in the best possible way. As an example of such a diagram, consider the one in Figure 6.5 of a satellite TV receiver. The signal is a composite consisting of audio and video. The block indicating the separation of these two signals is clearly marked (FM demod), and there should be no problem about their direction of movement and the circuits that will work on them. The unit is a double conversion receiver and so requires two mixer stages. One of the local oscillators is a fixed frequency type; the other is variable.

This block diagram does have an unusual feature—the inclusion of the frequencies handled by the various stages. Such information helps in an understanding of the unit.

COLOR TV RECEIVER BLOCK DIAGRAM

The block diagram in Figure 6.6 is that of a color TV set. Fortunately, the connecting lines are equipped with arrows (something not always done) and so the various signal paths are easier to identify. Further, each block is marked as to its function. Many block diagrams are not as detailed as this one.

Tracing paths in a color television diagram becomes easier by deciding on which of the three signals is to be followed since the overall diagram is a combination of a sound diagram, a video, and a sync circuit diagram. In addition to these there are two power supplies, one furnishing low-voltage DC and the other delivering high voltage for the picture tube.

With the exception of the picture tube, color TV receivers are all solid state, using ICs extensively. The composite video signal, consisting of frequency modulated (FM) audio and amplitude modulated (AM) video, is picked up by an antenna with each particular channel selected by a tuner, either manually or via a remote control unit.

No matter which channel is chosen, the composite video signal is heterodyned by a local oscillator, resulting in frequency downconversion. The resulting intermediate frequencies (IFs) are generally 41.25 MHz for sound, 45.75 MHz for video. Following demodulation, the audio signal may be brought into a pseudo stereo modulator, then into a preamp followed by a power amplifier, prior to delivery to a speaker system. Some TV receivers have provision for the use of headphones.

The video is further amplified in its IF amplifier circuitry and is then demodulated. The TV set may be equipped with a comb filter for signal improvement at its high-frequency end. The R-Y, B-Y, G-Y, and luminance signals are usually brought into video amplifiers and are then used to modulate the signal electrode of the picture tube.

Horizontal and vertical synchronizing (sync) pulses are present in the video IF amplifier circuitry. These are brought into a sync separator followed by a sync

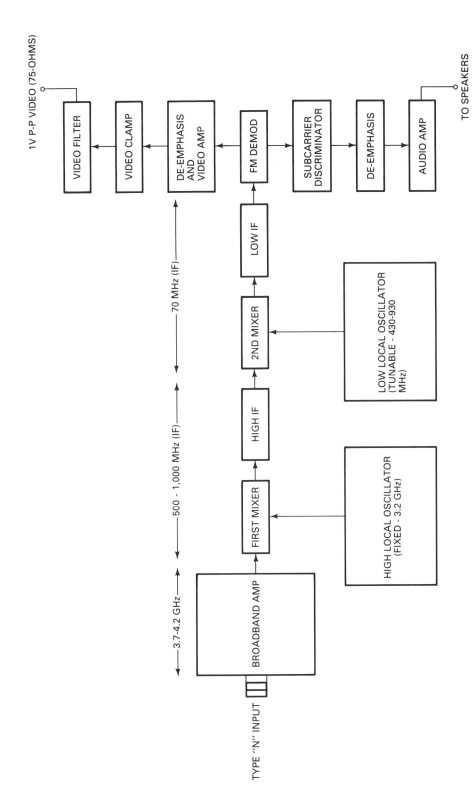

Figure 6.5. Double frequency conversion satellite TV receiver.

201

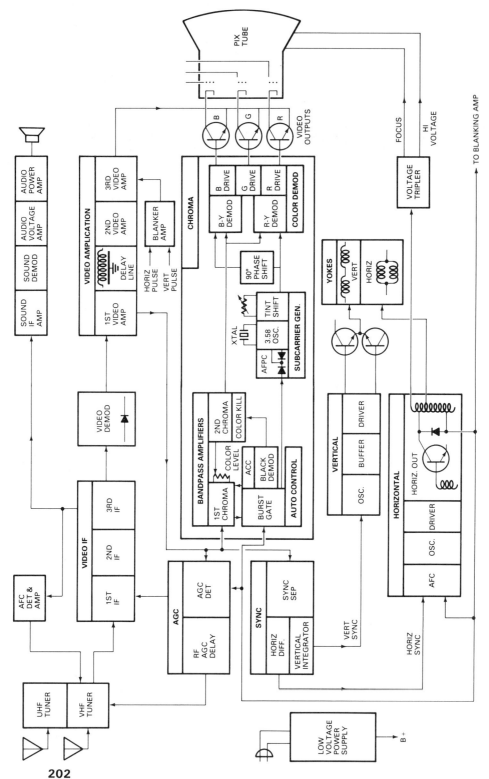

Figure 6.6. Color TV receiver block diagram.

amplifier/clipper circuit. The vertical and horizontal sync pulses are then separated. The vertical sync pulses feed a vertical integrator whose output controls the operation of a vertical oscillator. The vertical sync output is brought into a deflection yoke around the neck of the picture tube, and the vertical sync pulses then control the vertical sweep of the cathode-ray beam.

Following the sync amplifier, the horizontal sync pulses are used to control the frequency of a horizontal oscillator. The horizontal sync pulses are strengthened in a horizontal amplifier and are then delivered to horizontal sweep coils that form part of the deflection yoke. These control the horizontal sweep of the scanning beam in the picture tube with the scanning beam sweeping nearly 16,000 lines per second.

The high voltage required for the picture tube is obtained from the horizontal sweep circuitry. The horizontal amplifier operates a horizontal-output transformer, sometimes called a "flyback" transformer, for producing a high voltage at a frequency of approximately 15,750 Hz. This voltage is then rectified and connected to the high-voltage anode of the picture tube.

TV Block Diagram Variations

There are circuit differences between TV sets to such an extent that there is no such thing as a master diagram that could be used for all receivers. The diagram in Figure 6.7 is just one example. The power supply, for example, is an individual module, as indicated by the dashed lines, and it also contains a degaussing coil. The picture tube in a TV receiver can be affected by the earth's magnetic field or that produced by some electrical device. The purpose of a degaussing coil is to demagnetize the picture tube, by supplying a magnetic field produced by an alternating current. Degaussing coils are also available as separate units for those TV receivers not so equipped.

Still another feature is a comb filter module, enhancing picture detail. The receiver is also equipped with a pseudo-stereo module and so will either contain left and right sound speakers or will be capable of driving an external stereo system.

MONOCHROME TV RECEIVER BLOCK DIAGRAM

These TV sets (black and white pictures only) are simpler than color receivers since color circuitry isn't required. The picture tube is also not as sophisticated.

The monochrome TV set in Figure 6.8 is a composite consisting of blocks and a single IC. The dashed lines enclose the sound-demodulation transformer (T1) to indicate that this circuitry is shielded. A single IC containing a preamplifier is used, and it delivers its output to an audio-frequency (AF) amplifier. The outer dashed lines enclose the entire sound section with the exception of the AF amplifier and monophonic speaker.

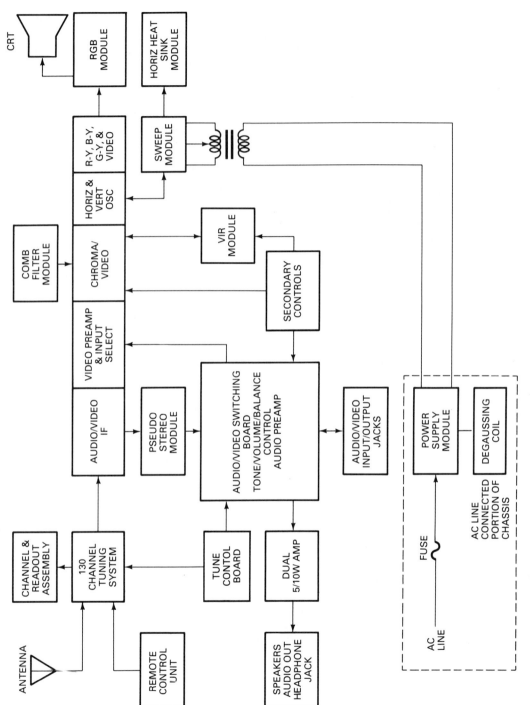

Figure 6.7. Block diagram of color TV receiver featuring simulated stereo, enhanced audio/video input, and output switching. (Courtesy General Electric).

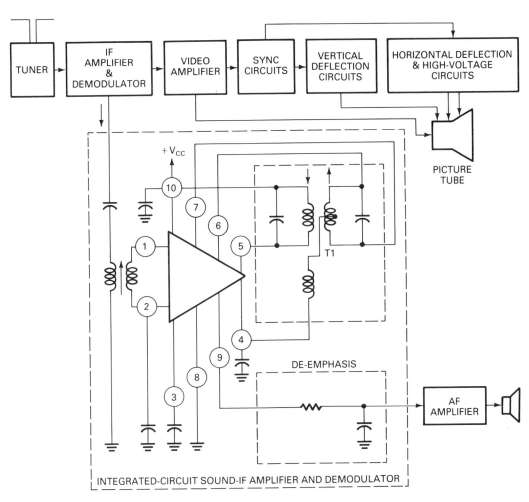

Figure 6.8. Use of IC simplifies circuit diagram.

VIDEO MONITOR

TV receivers are integated units that include a front end tuner, IF stages, and a video demodulator. A monitor, on the other hand, is a picture display unit (Figure 6.9) that does not contain these sections. A TV set has limited signal input possibilities consisting only of antenna signals. Further, no TV set has an output port with the possible exception of a headphone jack. While a monitor does not have an antenna terminal block, some have provision for as many as 23 signal inputs plus additional outputs for speakers. These inputs include jacks for connecting a VCR, videodisc player, video games, cable TV, and stereo audio signals, with the total number and variety of input and output ports varying from one monitor to another.

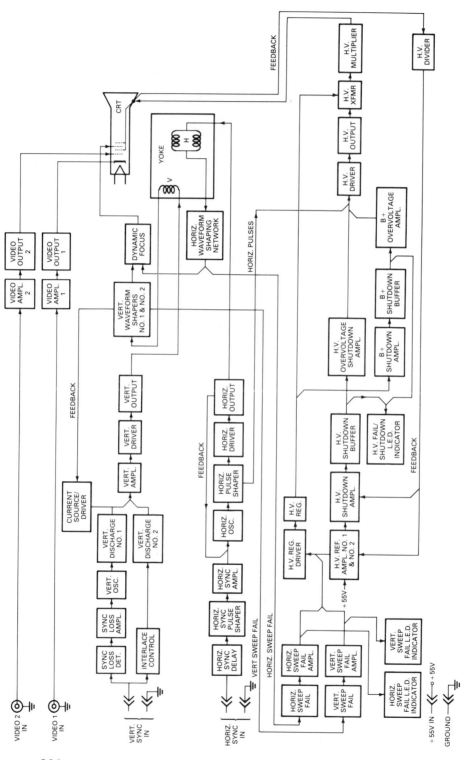

Figure 6.9. Block diagram of video monitor. (Courtesy Motorola, Inc.)

206

A TV receiver/monitor is a unit that combines the TV stages that are missing from a monitor, plus all the stages that comprise a monitor. Its advantage over an integated TV set is the larger number of input and output ports with which it is equipped.

TV Vertical Oscillator and Amplifier

The purpose of this oscillator is to help supply vertical sweep for the picture tube. Its free-running frequency is approximately 60 Hz. The output of this low-frequency generator is used to drive a vertical amplifier, with the current output of this circuit delivered to the vertical deflection coils positioned on the neck of the picture tube. The incoming video signal contains sync pulses used to trigger the vertical sweep oscillator keeping that oscillator operating at 60 Hz (Figure 6.10).

The base-emitter circuit of Q1 (Figure 6.10A) is reverse-biased, cutting off or blocking current flow from the emitter to the collector, from which the circuit derives its name of blocking oscillator. Diode D1 works as a pulse damper, acting as a short circuit for the momentary pulse of voltage caused by the collapsing magnetic field around the primary of transformer T1. The frequency of the oscillator can be changed within small limits by vertical hold control R2.

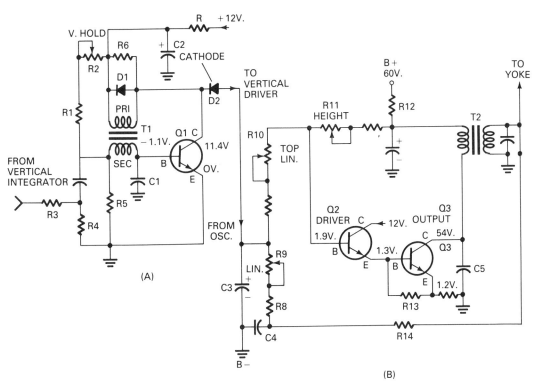

Figure 6.10.   Vertical sweep circuits. Vertical oscillator (A); vertical amplifier (B).

In the vertical amplifier circuit (Figure 6.10B), driver transistor Q2 is used as an emitter follower and is directly coupled to the base input of Q3. The circuit has three variable controls. There are two linearity controls, R9 and R10. The first of these adjusts overall vertical linearity, while R10 helps control linearity at the upper part of the picture. Picture height is adjusted by R11. These three resistors, plus capacitors C3 and C4, form a pulse-shaping network.

The output transistor, Q3, is a power type mounted on a heat sink. The amplified vertical sync pulses are sent to the vertical windings on the yoke mounted on the neck of the picture tube, through vertical output transformer T2.

Horizontal Automatic Frequency Control and Oscillator

Unlike the vertical oscillator, the horizontal oscillator (Figure 6.11) depends on some form of automatic frequency control (AFC). The reason for this is that noise pulses do not affect the vertical oscillator but can throw the horizontal oscillator off frequency.

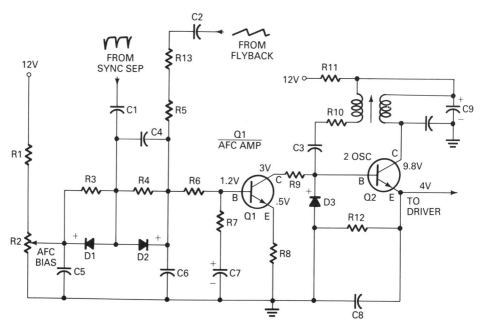

Figure 6.11.   Horizontal AFC and oscillator.

Two pulse voltages, sync and flyback, are combined in diodes D1 and D2, connected back to back. If the flyback pulses and sync pulses have the same phase and frequencies, the diode current makes these components behave like low-value resistors. They shunt resistors R3 and R4, effectively removing them from the circuit. Under these conditions, no correction voltage is applied to the input of Q1, and the horizontal oscillator continues running on this frequency, which happens, at the moment, to be identical with that of the TV sync pulses. If these pulses do not have the same frequency, a difference voltage will develop across R3 and R4. In turn, this voltage will change the operating frequency of the horizontal oscillator. Thus the AFC amplifier, Q1, controls the operating frequency of horizontal oscillator, Q2. The output of this transistor is emitter-coupled to a horizontal driver transistor, Q1 in Figure 6.12.

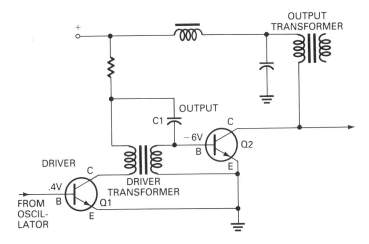

Figure 6.12.   Horizontal driver and output circuits.

Horizontal Output Circuit

The input to the horizontal driver, Q1 in Figure 6.12, consists of square-wave pulses that last for 22 microseconds of the total horizontal scan time of 63 microseconds.

The input from the preceding horizontal oscillator drives Q2 into conduction. Ordinarily, the output of the horizontal driver would also be a square-wave pulse, but capacitor C1 connected between the primary and secondary of the driver transformer changes the waveform from a square wave to a sawtooth. The combination of square-wave and sawtooth voltages produces a waveform called a "trapezoid." This wave, appearing across the output transformer, produces a sawtooth current through the transformer resulting in a linear sweep of the cathode-ray beam across the screen of the picture tube.

SURFACE ACOUSTIC WAVE FILTER

Abbreviated SAW filter, these are suitable for PC board mounting and are used to replace conventional LC inductor/capacitor filters (Figure 6.13). They do not require tuning and have a high insertion loss so they may be accompanied by amplifiers. These units are used in modulators, demodulators, converters, decoders, signal processors, and broadcast TV and satellite TV receivers.

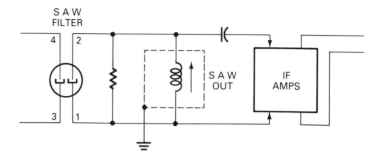

Figure 6.13. SAW filter circuitry in an IF stage.

SAW filters are manufactured on various piezoelectric substrates, commonly lithium niobate (LiNbO3) and ST-quartz. There are two major types: the SAW transversal and the SAW resonator. The transversal has excellent passband behavior and stopband rejection. The resonator is used for narrowband, low-loss filters in the 30 MHz to 1 GHz or more range. The use of a SAW filter, plus additional amplifiers, increases manufacturing cost but results in pictures that show less noise and are more interference free.

COMB FILTER

Because the chroma signal is contained within the space occupied by the luminance (black-and-white) signal, there is some luminance/chroma signal crossover resulting in indistinct picture resolution and fringing or cross-color interference.

These problems are eliminated through the use of a comb filter (Figure 6.14), a circuit arrangement that "combs out" color information while the chroma information surrounded by the luminance signal is able to pass through. This increases picture resolution by about 22 percent.

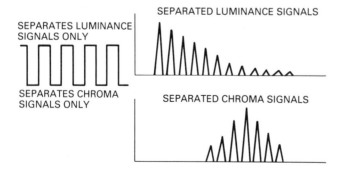

Figure 6.14. Comb filter produces more accurate separation of chroma and luminance signals.

VIDEO AMPLIFIERS

The amplifier in Figure 6.15 uses a pair of JFETs, with the input transistor's drain directly coupled to the gate of the following transistor. The source resistor for Q1 is bypassed by a large value electrolytic. L1, connected to the drain of Q1, is used to supply high-frequency peaking. Without this coil there would be a loss of high-frequency detail in the picture. The output of the video amplifier is connected to the signal input of the picture tube.

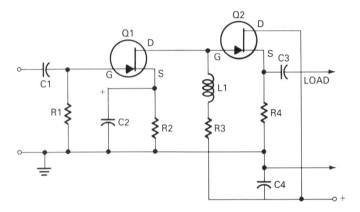

Figure 6.15. JFET video amplifier.

The function of the video amplifier is to supply an adequate amount of picture signal voltage so as to drive the input signal element of the picture tube satisfactorily. This element can be the cathode or control grid of that tube, depending on the phase of the output signal. Unlike the output amplifier in a receiver that must supply adequate power to operate a speaker system, the video amplifier is a voltage type.

The band of frequencies supplied by the video amplifier covers a much wider range. In the audio amplifier it extends from 20 Hz to 20 KHz, although some of these amplifiers have a much wider passband range than this. For a video amplfier the band to be amplified extends from 60 Hz to 3.5 MHz–4.2 MHz, with the latter figure preferable. Various phantom capacitances have the effect of bypassing the signal, with this effect increasing as frequency rises. Both series and shunt peaking coils are used to overcome capacitance effect.

The output of the video demodulator is fed into a bipolar transistor, a P-N-P in this illustration (Figure 6.16). Two signals are available at the output of transistor, Q1, an emitter follower. One of these is the audio signal, which is picked off the collector via L3, with the 4.5 MHz audio signal developed across it; the other is the video signal supplied by the emitter. The input to Q1 uses a pair of peaking coils, L1 and L2. The video signal is developed across R4, which not only acts as the video signal load but is also part of the biasing circuit for Q1. The input to Q1 is high impedance; the emitter is low impedance.

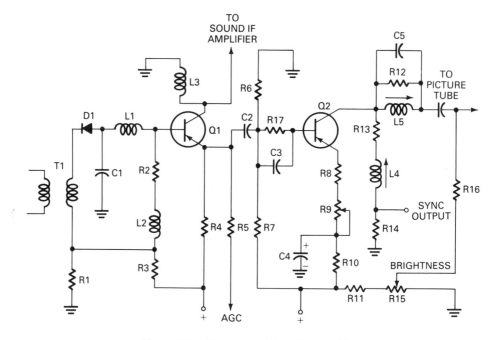

Figure 6.16.   Bipolar transistor video amplifier.

R17 shunted by C3 in series with the base of Q2 works as a high-pass filter and helps maintain a high-frequency response. The coupling capacitor, C2, has a high value of capacitance, usually about 10 μF, and sometimes more. This component has a double function: it blocks the application of the DC voltage applied to the emitter of Q1, and at the same time it has a fairly small reactance to the passage of low video frequencies. The output circuit of Q2 contains both series (L4) and shunt (L5) peaking coils. Sync output is obtained from a low-impedance point in the collector circuit of Q2, with this tap-off point far down on the collector load (R14) to prevent sync overdrive.

The video amplifier in Figure 6.16 also shows the use of a series diode (D1) working as a signal demodulator. The L-C network connected to the anode of this diode separates the combined audio/video signal from the carrier. This filter eliminates the carrier wave, while the baseband audio and video signals, no longer modulated onto the carrier, are passed on to Q1 for application.

The signal is capacitively coupled (C2) to Q2. The variable resistor (R9) in the emitter circuit of this transistor is a gain control.

Video IF Amplifier

The video IF amplifier (Figure 6.17), consisting of at least one stage but generally more, contributes substantially to the gain and selectivity of the TV receiver. The amplifier is a link between the front-end tuner from which it receives the IF signal

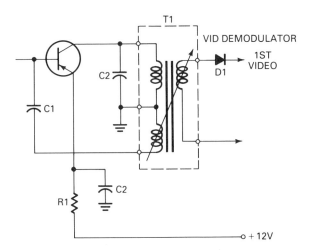

Figure 6.17. Video IF amplifier. This stage is the one immediately preceding the demodulator.

and the video demodulator to which it delivers the signal. AGC voltage picked up from circuitry following the video diode demodulator is fed back to one or more of the IF stages. The input IF stage is equipped with a pair of series-resonant traps, with one of these tuned to 47.25 MHz, the other to 41.25 MHz.

AM AND FM RECEIVER CIRCUITRY

Like a TV receiver, an AM or FM set is an integrated unit but simpler because it consists of just two sections: a tuner and an audio amplifier. In some home hi/fi systems these two are separated into individual components. The audio amplifier can be further subdivided into a pre- or voltage amplifier and a power amplifier. Although the tuner usually remains as a distinct unit, it is made up of a number of partial circuits consisting of a front end, a series of IF stages, and a demodulator (detector). The input to the tuner is a modulated signal supplied by an AM or FM broadcast station; the output is unmodulated audio to be supplied to the input of a preamplifier. Some receivers are combined AM/FM units, with each of these supplying monophonic (mono) sound only. A set that is identified as AM/FM/MPX (multiplex) can supply stereophonic (stereo) sound supplied by the FM section, although some do produce stereo for both AM and FM.

With the possible exception of a few specialized types, all radio receivers today are superheterodynes, more commonly referred to as "superhets." The broadcast signal is picked up by an external FM antenna or by a built-in loopstick for AM reception. Although not indicated as such in circuit diagams, the FM antenna is a circuit, broadly tuned to a wide band of frequencies. Tuning of the antenna is done physically and depends on the length to which the pickup element of the antenna is cut. The antenna signal is brought into the receiver via a length of transmission line, sometimes called a "download," consisting of either two-wire line (twin lead) or coaxial cable. Both of these use a pair of conductors for transferring the signal picked up by the antenna. For AM reception just a single wire is used.

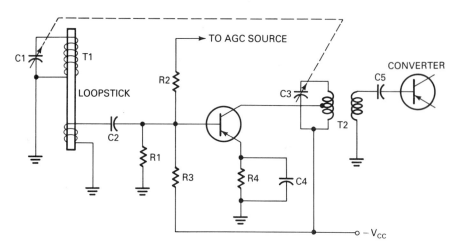

Figure 6.18.   RF amplifier.

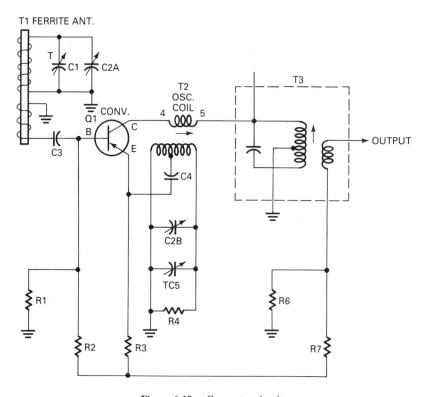

Figure 6.19.   Converter circuit.

CONVERTER

The AM or FM signals may first be brought into a tuned radio-frequency (RF) amplifier, a straightforward circuit (Figure 6.18) for strengthening the signal, although this stage may be omitted by some low-cost receivers. From this stage the signal is fed into a converter circuit (Figure 6.19), a combined mixer and local oscillator. (Figure 6.18 and Figure 6.19 are for AM receivers.) In some receivers the mixer and local oscillator are separate circuits. The function of the mixer and local oscillator, or the converter, is to change the carrier frequency of all broadcast station signals, whether AM or FM, to a single, fixed frequency, the intermediate frequency. This technique is referred to as "downconversion" since the intermediate frequency resulting from this process is always lower than the carrier frequency. In effect, the broadcast carrier is replaced by another carrier. This is the only modification that takes place, because the audio signal itself remains unchanged.

Depending on the design of the receiver there may be one or more IF stages, with the signal finally fed into a demodulator, sometimes called a "detector." The output of the detector is audio only, with the intermediate frequency stripped away and discarded.

THE DEMODULATOR

The demodulator can be either a diode or transistor with the diode the more commonly used component. The diode receives the modulated IF signal from the secondary winding of the final IF transformer (Figure 6.20). The signal is AC, hence is alternately plus and minus, with the frequency that of the IF. When the signal is negative it makes the cathode of the diode similarly negative, and current flows in the demodulator circuit. We can trace this current flow from the top of the secondary of the IF transformer (T1) to the cathode of the diode, which has now been made negative with respect to its anode. The current flows through the diode, through load resistor R1, back to the bottom end of the IF transformer secondary winding, and then through that winding.

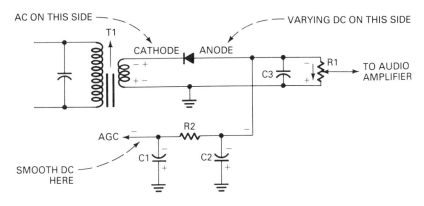

Figure 6.20. Demodulator circuit and AGC.

When the signal polarity reverses, the diode will no longer conduct, consequently the signal has been changed from AC to varying DC. It is DC because it flows in one direction only. At this time the signal, although unidirectional, consists of two parts: the intermediate frequency and the audio. The IF, no longer necessary, can be disposed of by simply putting a capacitor (C3) across potentiometer R1. Since this bypass capacitor has a reactance that varies inversely with frequency, it will bypass the much higher frequency of the IF, but will have a substantially higher opposition to the passage of audio frequencies. As a consequence, most of the audio signal passes through R1.

AUTOMATIC GAIN CONTROL

Tuning across a broadcast band can result in strong changes in volume caused by four factors: the distance to the station; the station's transmitting power; the sensitivity of the receiver; and the position of the receiving antenna with respect to the signal. One that is broadside to the broadcast antenna picks up a stronger signal.

Some of these factors cannot be eliminated and so the problem cannot be completely cured, but it can be modified somewhat through the use of an AGC (automatic gain control) circuit, shown associated with the diode demodulator of Figure 6.20. The voltage at the top end of R1 is varying DC, but this can be smoothed by an RC filter consisting of C1, C2, and R2. The output of this filter can be considered as a rather steady DC voltage, for a single, particular station. The voltage at the top end of R1, though, will vary in strength, depending on the strength of the signal presented to the diode. This means that the voltage at the output of the AGC filter will be higher for a strong signal than for a weak signal.

This DC voltage is fed back in the form of bias to one or more preceding IF stages and is of a polarity such that it opposes the existing bias, reducing the gain of the controlled stage. Unfortunately, this AGC system reduces the gain for weak as well as strong signals, but it does have some benefit since it reduces signal blasting as the receiver is tuned across the broadcast band.

Auxiliary AGC

There are two ways of removing the ineffectiveness of the AGC circuit of Figure 6.20. The first is to dispense with the AGC system altogether and depend entirely on the volume control for the level of sound output. The other is to use an auxiliay AGC circuit, one that can distinguish between signal levels that require AGC action and those that do not (Figure 6.21).

The demodulator diode remains but is not shown in this circuit. A second diode is used specifically for AGC action. The first IF stage isn't AGC controlled, but the second stage is. In the absence of a signal the collector currents of the two N-P-N transistors, Q1 and Q2, will be the same, assuming the use of identical transistor types. In the case of R1, collector current will move down through this resistor in the direction shown and with the polarity of the voltage drop across it as indicated.

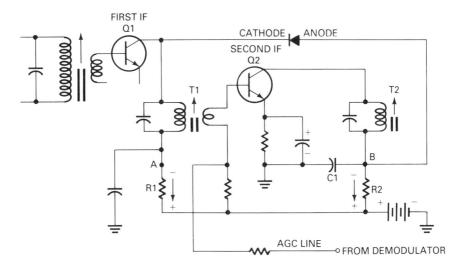

Figure 6.21. Auxiliary AGC system.

The same voltage having the polarity shown will also appear across R2, because this resistor is connected in the collector circuit of the second transistor. As a result, the voltage at point A will be the same as that at point B.

Move upward from point A through the primary winding of T1 to reach the cathode of the AGC diode. Do the same by moving upward from point B, arriving at the anode of this diode. Thus, the same voltage appears on the cathode and anode, and the diode does not conduct.

Assume that a weak or moderate signal is being received. It will be passed from the first IF stage to the second. This will result in a decrease in gain of this stage because of the AGC voltage received along the AGC line from the diode demodulator. In short, the existing AGC circuit is working. But only the second IF stage is affected because it is the only stage connected to the AGC line. As a result, the collector current through Q2 will be reduced. This collector current, as indicated earlier, flows through R2. With a reduced current through this resistor, point B becomes less negative. For weak or moderately strong signals, this change in potential level at point B is insignificant, hence the signal level of the second IF isn't affected.

For a strong signal, the AGC circuit will reduce the collector current of Q2 and so the voltage drop across R2 will be greatly reduced. Point B will become much less negative, but this point is connected to the anode of the auxiliary diode. As a result the diode goes into conduction. The strong signal, appearing at the collector of Q1, now has a path through the auxiliary diode. This bypassing action by the diode leads the signal to point B where it is bypassed to ground by capacitor C1. Of course not all of the signal follows this path, since that would lead to no sound output. Enough of the signal, though, is routed to ground so that tuning across the broadcast band does not result in sound blasting.

This does not mean all stations will sound weak. The volume control can be turned up so that weak stations will be stronger, but strong signals will have their level reduced.

DECOUPLING CIRCUITS

Transistor receivers have a common element—the power supply. The power supply, whether battery or AC operated, represents a common impedance, linking all the circuits in the receiver. For AC-operated power supplies this does not present a problem since the capacitors used in the power supply filter have a large amount of capacitance, hence a very low impedance.

When the DC voltage source consists of batteries, the problem of a common impedance can be troublesome. When a battery is fresh or fully charged, its impedance is very low, but as the battery gets older or discharged, it impedance increases. This can result in unwanted coupling between the various stages of the transistor receiver, resulting in erratic behavior.

There are several ways of overcoming the difficulty. The most obvious solution would be to replace the batteries with new, fresh ones. However, this could mean putting in new batteries before the old ones have reached the end of their useful life. Another technique is to use the same idea employed in filters used by AC power supplies.

The filter (Figure 6.22) is an RC type, consisting of R1 and C1. These can be used by all the stages, or possibly just by those stages affected by the common impedance of the batteries. Another technique is to shunt the battery supply with an electrolytic capacitor. In this case, the higher the capacitance of this bypass the better.

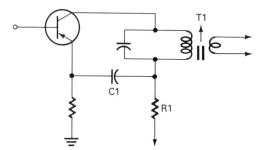

Figure 6.22. R1-C1 decoupling circuit.

CIRCUIT VARIATIONS

In some instances, a partial circuit composed of a number of subcircuits will have some modifications or changes, and the reason for these may not be immediately apparent. One of these is shown in Figure 6.23. Here, the collector of the N-P-N transistor isn't connected to the top (point A) of the IF transformer, as might be expected, but is tapped down; that is, it is connected to some point near the lower end of the coil winding.

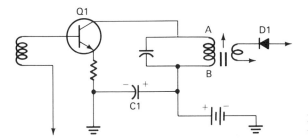

Figure 6.23. Tapped down IF transformer.

We get a maximum transfer of signal energy when the impedance of the source and that of the load are the same. The maximum impedance of the primary winding of the IF transformer in this diagram is at point A, but is less as we tap down. The impedance of the collector in this example (the source) is somewhat lower than the impedance at point A. Consequently, to match impedances, a connection is made to some point lower down on the coil.

Not only the primary winding of an IF transformer, but the secondary can be tapped down (Figure 6.24). In this example the secondary of the input IF transformer is tapped down to match the input impedance of the N-P-N transistor, that is, the impedance between the base and the emitter. The primary of the output IF transformer is also tapped down to match the impedance between the collector and emitter. The location of the tapped points depends on two factors: the overall impedance of the coil winding and the amount of impedance required.

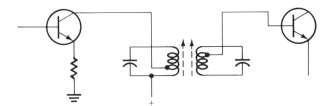

Figure 6.24. Both sides of IF transformer are tapped down.

NEUTRALIZATION

One problem facing IF amplifiers, and other types of amplifiers as well, is that the output is greater than the input. If some portion of this output is permitted to leak back to the input so it is in phase with the input signal, two effects may take place, both undesirable.

If there is a relatively small amount of feedback, the gain of the affected stage will increase and the bandpass will decrease. With a sufficient amount of this kind of feedback, known as "positive" or "regenerative" feedback, the circuit will no longer be an amplifier but an oscillator, and as such will generate its own signal.

In the first case, the signal will sound much sharper and will be unstable; in the second the circuit working as an oscillator may transmit its own signal, possibly producing a whistle in the received signal and equally possibly interfering with signals being received by outside, nearby radio sets.

To counteract positive feedback, some portion of the output signal is fed back to the input, but out of phase with the incoming signal. One method of supplying out-of-phase feedback, also called "negative" feedback, is indicated in Figure 6.25. Feedback takes place here with the help of capacitor C1 connected between the bottom of the primary winding of the IF output transformer (T2) to the top of the secondary winding of IF input transformer (T1). The amount of feedback can be controlled by careful selection of the value of capacitance of C1. The larger its capacitance, the greater the amount of feedback.

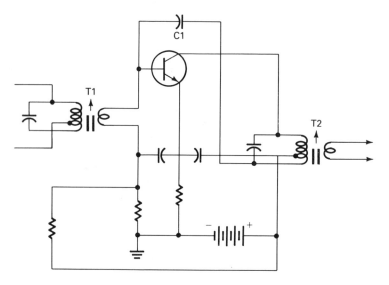

Figure 6.25. Capacitor C1 supplies negative feedback.

Offhand it might seem that the larger the amount of negative feedback the better. The problem, though, is that it reduces gain. This takes place since the feedback signal is out of phase with the input signal, reducing the amplitude of that signal.

DRIVER AMPLIFIER

A driver amplifier, also known as a voltage amplifier, supplies the signal driving voltage for a following power amplifier (Figure 6.26). Driver stages generally consist of a single transistor, operated class A. The input to the driver can be any unmodulated audio signal supplied by the demodulator output of a receiver, or a microphone, or the audio output of a compact disc player.

In this circuit, fixed bias is supplied by R1 and R2 shunted across the DC supply, B1. Self-bias is obtained from the voltage developed across R3 with this voltage smoothed by C2. This circuit uses capacitive coupling from the preceding stage via C1. This capacitor not only supplies signal transfer but blocks any DC supplied by the earlier stage from reaching transistor Q1.

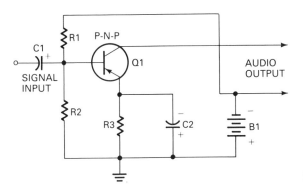

Figure 6.26. Driver amplifier.

TONE CONTROLS

The driver amplifier may make use of a tone-control subcircuit such as either one of the two in Figure 6.27. In Figure 6.27A the tone control is a subcircuit R-C network consisting of a fixed capacitor C3 and a variable resistor R4, a potentiometer. When the variable resistor is in its minimum resistance position, the capacitor effectively bypasses treble tones. This supplies a higher ratio of bass to treble, and so the output has bass emphasis. At the opposite end of travel of the potentiometer arm, there is maximum resistance in series with the capacitor, with less bypassing action, and so there is a greater ratio of treble to bass in the output.

Figure 6.27B is the same as A except that the potentiometer has been replaced by a rotary tapped switch. One of the connections is a solid line, indicating a direct connection to ground. With the switch in this position, the shunting capacitor is fully effective and has substantial treble tone bypassing action. As a result, bass tones sound emphasized.

BLOCK AND SYMBOL PARTIAL DIAGRAMS

It is sometimes convenient to combine a block diagram in a circuit using symbols (Figure 6.28). What we have here is an amplifier with a transistor circuit substituting for one of the bias resistors. The purpose of the block is to save time in drawing a complete circuit when it isn't necessary to do so. In this case we might be interested only in the details of the amplifier and might also want to show the relationship of the driver but without discussing that driver further.

PHASE

Two signal voltages may start at the same time, reach positive and negative peaks at the same time, and complete their cycles simultaneously. These two voltages are in phase since they are always in step. An exactly opposite condition is obtained if the two signal voltages start at the same time but move in opposite directions and

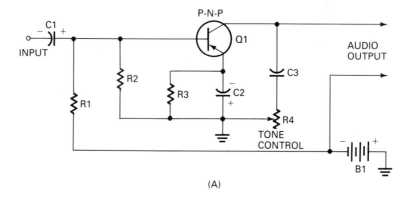

(A)

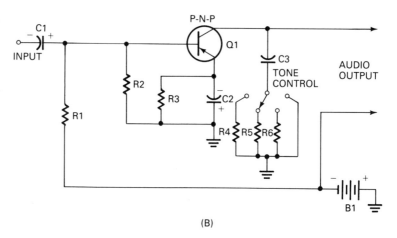

(B)

Figure 6.27. Tone control circuits. Potentiometer type (A); switched control (B).

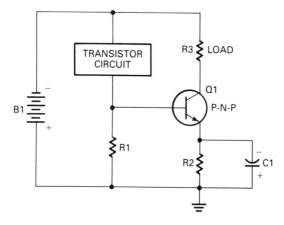

Figure 6.28. Transistor circuit substituted for one of the forward biasing voltage divider resistors shown as a block diagram within a circuit diagram.

so are out of phase. The maximum out-of-phase condition is 180°, but any lesser out-of-phase condition is also possible.

A characteristic of the grounded-emitter circuit is that it inverts the phase of the input signal (Figure 6.29). The input signal is supplied to resistor R1. The output appears across R2. If the signal voltage at a particular moment is such that the top end of R1 is negative and the bottom end positive we will have the equivalent of a voltage in series aiding with the bias battery. When this happens more current will flow through the transistor from the collector to the emitter. This current will move through R2 making point C more negative with respect to point D. Thus, as the input signal becomes more positive, the output at C becomes more negative. When the opposite condition occurs, when point A becomes more negative, point C will become more positive. Thus, this transistor not only amplifies the input signal, but inverts it in phase as well. Neither the grounded collector nor the grounded base supply signal phase inversion. One of the factors, then, determining the use of a particular amplifier circuit is whether we want phase inversion of the signal or not.

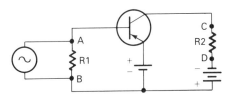

Figure 6.29. Output of a grounded-emitter circuit is out of phase with input signal.

AMPLIFIER CLASSIFICATION

The arrangement of solid-state amplifiers into grounded emitter, grounded collector, and grounded base is just one that is possible. Another method is to classify them as voltage or power amplifiers, and still another by class, such as class A, class B, and so on. Or we can categorize them as single-ended or push-pull. In some instances several of these identifying names are used.

CLASS A, SINGLE-ENDED, POWER AMPLIFIER

This amplifier (Figure 6.30) uses three names to indicate its type. A class A amplifier has limited efficiency but supplies amplification with minimum distortion. Single-ended means that a single transistor is used for delivering audio power to a following speaker system. Power amplifier means that it is the final transistor prior to the speaker and is used for delivering audio power to it.

The audio signal is supplied by a prior stage and is called a preamplifier or sometimes simply a driver. The transistors used by both the driver and power amplifier are P-N-P types. This does not mean both transistors are identical, because the output transistor, Q2, is a power type designed to handle relatively large amounts of current. During operation it can get quite hot and so may be mounted on a heat sink, a large metal plate used for radiating heat.

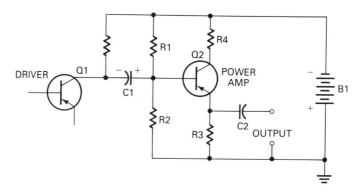

Figure 6.30. Single-ended power amplifier.

PUSH-PULL AMPLIFIER

The power amplifier in Figure 6.31A is a push-pull type, operated class B. The circuit can be redrawn as in Figure 6.31B to show it is really a pair of single-ended units with bias supplied by R1 and R2 with these determining the operating point of the two P-N-P transistors. The battery also supplies the DC operating potential. The input signal is developed across the secondary of the input transformer and appears between the emitter and base of each of the transistors. This signal voltage is AC, and so the top and bottom of the secondary winding are alternately plus and minus. If the top of the winding is plus, transistor Q1 is driven into cutoff. At the same time the bottom of the transformer is minus, and so transistor Q2 conducts. When the input signal changes polarity, Q2 is cutoff and Q1 conducts. Thus, each transistor takes its turn in supplying signal current to the output.

Class B operation has a higher efficiency than class A. The single-ended transistor stage shown earlier in Figure 6.30 can only operate class A, hence does not have the efficiency of the class B arrangement.

Hybrids

Interstage and audio output transformers are generally found in older receivers because they have been replaced by direct or capacitive coupling. Some receivers have retained the interstage transformer with direct coupling to the speaker system, while others have eliminated the interstage but have retained the output transformer. Such receivers are referred to as hybrids. The word "hybrid" is used in other applications as well. Every TV set is a hybrid since it uses semiconductors throughout but with a single vacuum tube—the picture tube.

PUSH-PULL FOR RF

Push-pull circuits are applicable for either audio- or radio-frequency (RF) applications. Those in Figure 6.32 are intended for RF. Figure 6.32A shows a pair of N-P-N

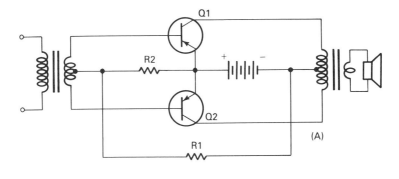

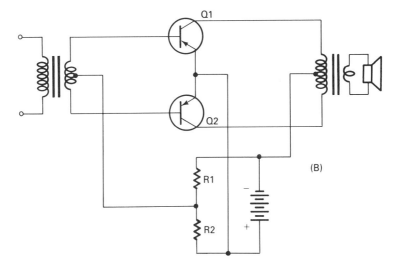

Figure 6.31. Push-pull amplifier. Class of operation is determined by amount of bias. Drawing A shows how circuit is usually drawn. Drawing B makes biasing and collector voltage more apparent. This circuit is generally found in older receivers as transformers have been replaced by other types of coupling.

transistors working in push-pull with the input signal supplied by an RF transformer. Self-bias for both transistors is via a single resistor, R1. The biasing is such that each transistor is at cutoff and is taken out of cutoff only by a signal of the correct polarity. When one transistor conducts, the other does not.

The output is also via a transformer with the primary winding tuned to the frequency of the input signal. The arrangement of this circuit is such that even-order harmonics of the signal are cancelled, but odd harmonics are reinforced. The circuit can not only be used as an amplifier of a fundamental frequency, but can work as a frequency tripler if the output transformer is tuned to the third harmonic. Although N-P-N transistors are shown in this diagram, it will work equally well with P-N-P

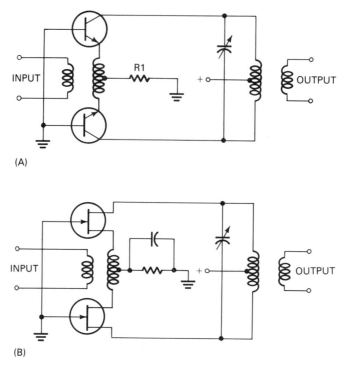

Figure 6.32.   Push-pull amplifiers for RF. Circuit using bipolar transistors (A) or FETs (B).

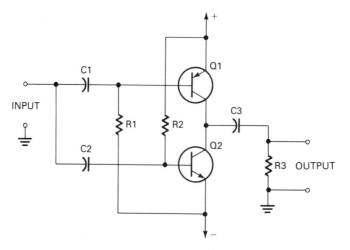

Figure 6.33.   Complementary symmetry arrangement.

types. In a circuit of this kind it is helpful if the transistors are as evenly matched as possible. Field-effect transistors can be used in place of the bipolar units (Figure 6.32B). The circuit remains the same except for the capacitor shunted across the biasing resistor.

COMPLEMENTARY SYMMETRY

Depending on the polarity of an input signal, it will either increase or decrease the forward bias of the transistors to which it is applied. But if the signal drives the base-emitter input of two different transistor types, such as the N-P-N and P-N-P in Figure 6.33, it will increase the forward bias of one and decrease the forward bias of the other during one-half of the input cycle and will produce the opposite effect during the other half of that cycle. Using this technique, known as complementary symmetry, eliminates the need for an input transformer.

TRANSFORMERLESS CIRCUITRY

Figure 6.34 illustrates four circuits that eliminate both input and output transformers. Each of the circuits is a complementary symmetry arrangement using P-N-P and N-P-N transistors. Circuits (A), (B), and (C) are direct-coupled to the output, while circuit (D) uses capacitor coupling.

DIODE COMPENSATED AMPLIFIER

In Figure 6.35A resistors R1 and R2 form a voltage divider across the battery supply. The current flow through these two resistors is such that the base is made negative with respect to the emitter with the amount of bias voltage determined by the respective values of R1 and R2. The bias will remain constant, as it should, provided R1 and R2 do not change values. Resistors, though, have a positive temperature coefficient and so their resistance will increase with a rise in temperature.

The problem can be overcome by using a diode for R2, as in (B). If the temperature should increase, the resistance between the base and emitter of the transistor will decrease. The reason for this is that transistors have a negative temperature coefficient, opposite that of resistors.

With an increase in temperature, more current will flow between the base and emitter. These two elements form a diode, but it is now shunted by an external diode (Figure 6.35B). Assume an increase in temperature. The resistance between the base and emitter will go down, resulting in an increase in current between the two elements. But the shunting diode will also have a decrease in resistance, and since it is in parallel with the base-emitter resistance, will supply a current bypass path. The use of a diode as a protection against temperature changes is known as diode compensation or stabilization.

Diode stabilization can be used in power amplifier circuits as well as in drivers. Actually, it is used more often in power amplifiers since the transistors in such circuits get quite hot.

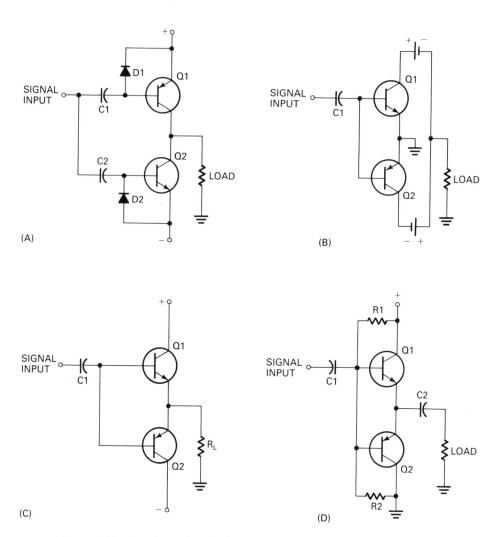

Figure 6.34. Transformerless circuitry. Common-emitter amplifiers (A and B); comon-collector amplifiers (C and D).

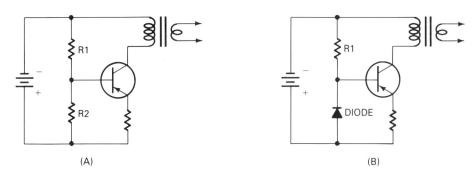

Figure 6.35.   R1 and R2 are used for forward bias (A). Diode substituted for R2 (B).

7

Logic Circuits

Logic circuits are solid-state devices using transistors or diodes, with these working as switches. Logic is sometimes referred to as symbolic logic or sometimes as digital logic since the switches are most often operated by pulses rather than analog waveforms. These circuits, also known as gates, are generally shown using special symbols. Their descriptive names are obtained from the way in which they function.

A large number of electronic switches can be used in components, and while it is easy to draw just one or two switches in a circuit diagram, the problem becomes quite complex when a large number are involved, particularly when the switches are interdependent.

There are two ways of designing and drawing switching circuits. The first, and the more difficult, is to plan a switching setup and then to breadboard it to see if it will work. This may be satisfactory if just a few switches are involved but becomes increasingly impractical as the number of switches is increased. The difficulty is that modifying the functioning or the location of a single switch out of a large group can affect the functioning of all the others, and so it can be time-consuming with no assurance of success.

A preferable method is to use a mathematical system, devised by George Boole in 1854, which can describe switching circuits, indicate the way in which they will interact, and enable the circuit designer to use the minimum number of switches.

Known as Boolean algebra, it was devised many years before the word "electronics" and its subject material became known.

AND CIRCUIT

One of the simplest of the logic circuits is the AND arrangement. It is shown in Figure 7.1 as a pair of series wired single-pole, single-throw (SPST) switches, coded SW1 and SW2, connected to a light bulb representing a load, and a voltage source E. The switches use the symbols for mechanical types, and although mechanical switches could be used, they represent those that are electronic for they consist of diodes or transistors. A pair of diodes or transistors, biased at or beyond cutoff, function as open switches. When the bias is lowered or removed, the diodes or transistors conduct. The important characteristic of these semiconductors is that they are either on or off, and there are no intermediate positions.

There are four possible operating conditions for this particular AND circuit. Switches SW1 and SW2 can both be on, that is, they are closed; SW1 can be on while SW2 is off (open); SW1 can be off when SW2 is on; SW1 and SW2 can both be off. This circuit is also called a logic gate, or a coincidence circuit, since it requires two (or more) simultaneous or coincident input pulses to produce a suitable output.

Figure 7.1. AND circuit represented by mechanical switches.

The semiconductor circuit (Figure 7.2) shows a pair of P-N-P transistors, but although Q1 is indicated as preceding Q2 it does not drive that transistor, and both Q1 and Q2 operate independently. Since there is no provision for forward biasing there is practically no current flow through the transistors from collector to emitter, and as a consequence practically no current flows through R1, the output load resistor. The voltage across this resistor is either zero or negligible, and a measurement taken

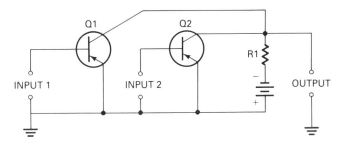

Figure 7.2. This basic AND circuit requires two coincident inputs.

across the collector and emitter would show it to be equal to the full battery voltage.

If both inputs receive a negative pulse—one that will make the base of each transistor negative and the emitter positive, the effect will be that of forward biasing, and both transistors will move rapidly into conduction, with the sum of the currents of the transistors flowing through common load resistor R1. There will be a voltage drop across this resistor, and the voltage between collector and emitter will decrease.

The output terminals are connected between the collector and emitter of Q2. In the absence of input pulse voltages, the voltage at the output is high. When simultaneous pulses reach input 1 and input 2, the voltage at the output will take a sharp drop. This decrease in voltage will last as long as the two pulses remain at inputs 1 and 2. When these pulses disappear, the output voltage rises to its maximum again. At the output, then, is a pulse voltage that could be used to trigger some other circuit.

In this circuit, if only one input pulse is received, at either input 1 or 2, the circuit action is ineffective. The flow of current through one of the transistors will produce a voltage drop across the load resistor, but the amplitude of this voltage will not be enough to actuate a following circuit.

TRUTH TABLE

While all the possible operating conditions of the semiconductor switches in the two-switch AND circuit can be described, a quicker and easier method is to use a truth table (Figure 7.3). The letters A and B at the head of the table refer to the switches; C is the output. In some truth tables the letter C is omitted and the word "output" is used instead. However, using a letter to represent output is the preferred form. The digit 0 is used as a symbol for an open switch; digit 1 represents a closed switch.

TRUTH TABLE

A B	C
0 0	0
0 1	1
1 0	1
1 1	1

(A)

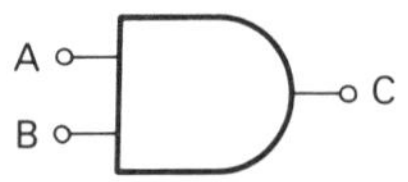

(B)

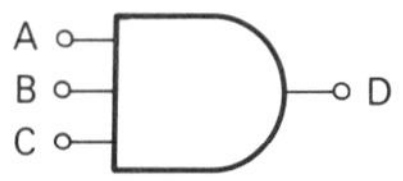

(C)

TRUTH TABLE

ABC	D
000	0
001	0
010	0
100	0
011	0
101	0
111	1

(D)

Figure 7.3. Truth table (A); AND gate symbol (B). Three input AND symbol (C); truth table for three input AND gate (D).

Digit 1 for a switch is sometimes called an "on" or "true" condition; digit 0 is an "off" or "false" condition.

The truth table is a way of indicating just how the AND circuit works. When switches 1 and 2 are open, that is, A AND B = 0, the output (C) is zero. When one of the switches is open (condition 0) the output remains at zero. Similarly, for a condition of 1 0, the output is still zero. However, when A AND B = 1 that is, when both switches are closed, we have output at C, indicated by 1 in the truth table.

The AND Symbol

The electronic symbol for a two-switch AND gate circuit is shown in Figure 7.3B. The letter A represents one of the switches; the letter B the other; while C is the output or load.

Gate Multiplicity

Although Figure 7.2 shows just a pair of switches, any number can be used. If, for example, the circuit includes three switches, the symbol representing the AND circuit would have three leads at the left of the symbol, marked A, B, and C, while the output could be identified by the letter D (Figure 7.3C).

Not only the symbol, but the truth table (Figure 7.3D) would be changed to indicate the presence of an additional switch. In this circuit all three switches must be closed for a true condition. If one or more of the switches are open, we would then have a false condition.

ALGEBRAIC REPRESENTATION OF GATES

An AND circuit consisting of just two or three switches is simple enough to draw, but if a larger number of switches are to be used it can be a time-consuming nuisance. An easier way is to represent each of the switches by a letter and to use a symbol to indicate the word AND. One of the available symbols used for this is an inverted V. To write switch No. 1 AND switch No. 2 we would then have $A \wedge B$. For a series switching circuit consisting of two switches A and B and a load C we could have:

$$A \wedge B = C$$

This Boolean algebraic equation tells us two things. The first is that we are working with a two-switch circuit. The next is that Switch A AND B must both be closed to be able to deliver a current, or possibly a signal voltage, to a circuit or a load, which we call C.

The same concept can be used for more than two switches. Thus we can have:

$$A \wedge B \wedge C = D$$

Each of the letters represents a switch, while the new letter D is the load. We now have a three-switch series circuit, and all of the switches must be closed for a true condition to exist.

Alternate AND Symbols

The inverted V is just one of several possible symbols used to represent AND but it is preferable since it is definite and precludes error. One of the other possibilities is to put the letters representing the switches immediately adjacent. Thus the letters AB would mean *A* AND *B*. The problem is that a pair (or more) of letters positioned this way means *A* multiplied by *B* in algebra and so its use in logic symbolism could result in some confusion. In algebra *ABC* is *A* times *B* times *C*. In symbolic logic it means *A* AND *B* AND C.

Another symbol to indicate AND is a dot (sometimes called a "connective") positioned halfway up between a pair of letters, as in *A* · *B*. The difficulty with this approach is that the dot looks like a decimal point or a period and unless correctly placed could be confused for either one of these.

DIODE AND CIRCUIT

A pair (or more) of diodes can be connected in series so as to function as an AND circuit (Figure 7.4). The diodes can be biased so they are in a current cutoff condition but on the simultaneous application of pulses of the correct polarity will drive the diodes into conduction. The disadvantage of the diodes, as opposed to transistors, is they do not supply signal gain. The diodes, however, can be followed by an amplifier, either with or without signal inversion. Diode AND circuits have been mostly replaced by transistors.

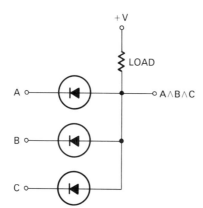

Figure 7.4. Diode AND circuit with three inputs. The truth table is the same as that for three transistors.

OR GATES

Figure 7.5A shows the basic concept of the OR gate. There are two switches, SW1 and SW2, in the circuit powered by a battery E and using a light bulb as the load. If either SW1 OR SW2 are in the on position, current will flow from the battery, through the closed switch, and through the light bulb. The truth table (B) adjoining

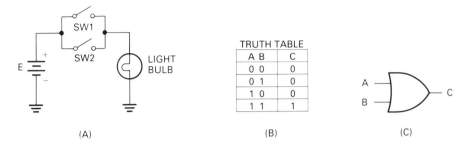

Figure 7.5. OR gate circuit using mechanical switches (A); truth table (B); OR symbol for two-signal input (C)

the diagram shows the four possible operating conditions. If both switches are open (0 or false) the output is 0 or false. If either switch, or both, are closed (1 or true) the output is 1. The same drawing shows the symbol (C) for an OR gate. The symbol in the illustration is for an OR circuit having two switches, but any number could be used and could be lettered *A*, *B*, *C*, and so on. The output can also have any letter.

Figure 7.6 shows the circuit diagram for an OR gate using a pair of N-P-N transistors, V1 and V2. Both transistors are in shunt since the collectors are tied together as are the emitters.

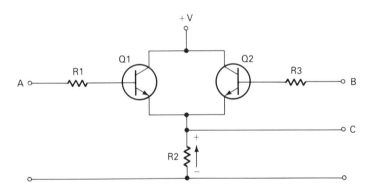

Figure 7.6. OR circuit. Inputs are at A and B. Output is across load resistor R2.

A 1 input (a true input) at either point A or point B will result in a 1 output (a true output) at point C. When either one or the other of the transistors, or both, are turned on, current flows through load resistor R2. This produces a voltage drop across this resistor such that the bottom end of the resistor is negative with respect to its top end as indicated by the arrow. But point C is wired to the top end of this resistor and so we get a rise in voltage at C. This output can be considered as a 1 or true voltage.

More transistors can be used in the OR circuit (Figure 7.7). In this case the transistors are P-N-Ps, and the collectors and emitters of each are wired in parallel. The load resistor is in the collector circuit instead of the emitter as indicated in Figure

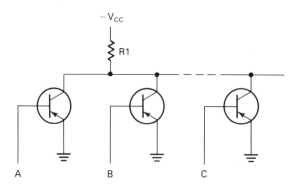

Figure 7.7. Triple input OR circuit.

7.6. The output is taken from the bottom end of R1. This circuit will deliver output if any one or all of the transistors conduct.

SYMBOLIC REPRESENTATION

There is only a certain amount of information that can be extracted from symbols used for logic circuits. The symbol can inform us of the type of circuit being used, that is, whether it is an AND or OR, and also the number of inputs. But we do not know if the semiconductors being used are diodes or transistors, and if transistors, the type being used. Further, the gate may be followed directly by a transistor amplifier so as to strengthen the output signal.

AND gates and OR gates are prepared on chips and offered in the form of integrated circuits. The IC can contain just a single gate, or a number of them, and are not restricted to logic gates of a single kind. To show the number of gates that are included in an IC they are grouped and surrounded by a rectangle, as in Figure 7.8. Each of the four OR gates shown here has its own inputs and output and, further, each of these gates can be operated singly or can be externally connected to work as a group.

The IC in Figure 7.8 has a dual inline package (DIP), but a single inline package

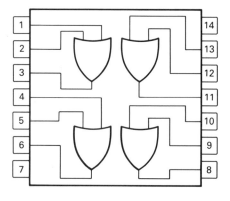

Figure 7.8. Dual inline package containing four OR ICs.

(SIP) can also be used. The DIP is often preferred since it allows for more spacing between the terminals used on the IC package.

EXCLUSIVE OR GATE

There are two types of OR gates: inclusive or exclusive. Unless otherwise indicated, an OR gate is inclusive. The inclusive OR, as previously described, supplies output if any of the inputs are true, representing a closed switch. With an exclusive OR the output is zero if both inputs are activated or if both are not activated. Figure 7.9 shows the symbol (A) and the truth table (B).

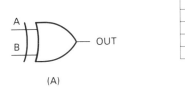

A	B	OUT
0	0	0
0	1	1
1	0	1
1	1	0

(A)   (B)

Figure 7.9. Exclusive OR gate (A) and truth table (B).

ALGEBRAIC REPRESENTATION OF OR GATES

The symbol V or the word OR always indicates a parallel circuit. Note that this symbol is exactly opposite that used for an AND arrangement. A statement such as $A \vee B$ means we have two switches, two because we have two identifying letters with the switches in shunt since we have used the OR symbol, V. The same symbolism can be used for any number of OR gates. For three gates, for example, we would have $A \vee B \vee C$. The complete Boolean algebra expression for three parallel gates would be:

$$A \vee B \vee C = F$$

In this example each switch is labeled A, B, C while the output is F. There is no standardization on switch lettering—any letters can be used.

Just as in the case of AND gates we have available a different symbol that can represent OR. This symbol is a + sign, and so $A + B$ means A OR B. The problem with using this symbol is that through education and experience we have come to associate it with addition, and in the Boolean algebra context it has a completely different meaning.

THE AND-OR CIRCUIT

While an AND gate or an OR gate can be used alone, quite often these are subcircuits and are part of a larger group of switches. Thus, it is possible to have a series-parallel circuit (Figure 7.10) possibly consisting of two gates in series and two gates in parallel. For a pulse to travel from the input, A, to the output, B, switches P AND

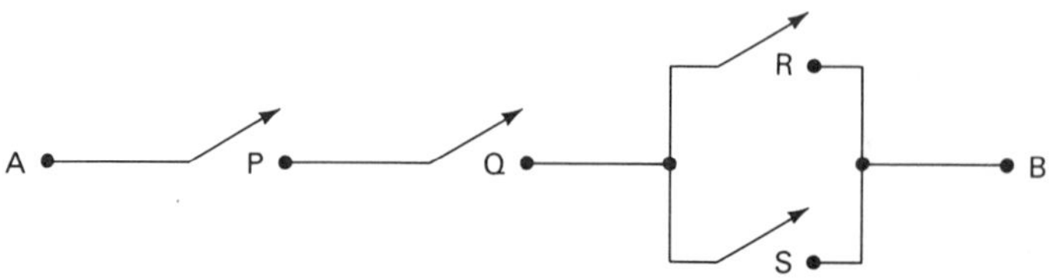

Figure 7.10. AND gate combined with an OR gate.

Q must both be in their closed condition (1 or true), and either switch R OR S must also be in its 1 state.

There are several optional ways of describing this circuit. It could be drawn by using either an actual circuit arrangement consisting of transistors, or one with equivalent mechanical switches as in Figure 7.10. A far simpler method would be to write a Boolean algebra statement.

$$(P \wedge Q) \wedge (R \vee S) = B$$

This statement supplies data concerning the gates in a precise, condensed manner. We know there are four switches, that the first two switches, P, Q, are AND connected, that the next two switches, R, S, form an OR subcircuit, and that the output is B.

Note the use of parentheses in the Boolean algebra statement of the circuit. Without them the statement would be capable of misinterpretation and would look like this:

$$P \wedge Q \wedge R \vee S = B$$

Consequently it is not possible to apply the statement in this form. However, with the statement arranged in parentheses it can be used for reducing the number of gates to a minimum.

CIRCUIT SIMPLIFICATION

The fewer the number of circuits in a component the lower its manufacturing cost and the less chance there is of a circuit fault. As an example, consider the series-parallel gates in Figure 7.11. Here a pair of series gates, P AND Q, are in parallel with another pair of series gates, P AND Q. The same letters are used for these gates because they are identical.

The Boolean statement for this circuit is:

$$(P \wedge Q) \vee (P \wedge Q) = B$$

We can read this statement as P AND Q OR P AND Q, but since both P gates are identical and both Q gates are also identical, we can simplify the Boolean expression to:

$$P \wedge Q \vee (Q \vee Q) = B$$

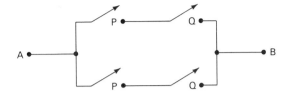

Figure 7.11. Four gates arranged in series-parallel.

We can read this statement and it will be exactly as the one written above: *P* AND *Q* OR *P* AND *Q*. In the simplified statement, though, there is just one gate, *P*. We can then redraw the gating circuit as shown in Figure 7.12. To avoid confusion one of the parallel gates can be relettered as *R*. The Boolean statement will become:

$$P \wedge (Q \vee R) = B$$

In the component in which this gating circuit is to be included we can then use the diagram of either Figure 7.12 or Figure 7.11. Since the two circuits can be made

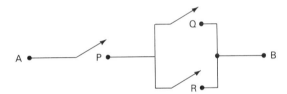

Figure 7.12. Simplification of the circuit in Figure 7.11.

to perform equivalent functions, the Boolean statements describing these gates are also equivalent. To show this equivalency we can write:

$$(P \wedge Q) \vee (P \wedge Q) \equiv P \wedge (Q \vee Q) = B$$

The gating circuits shown in the previous illustrations are fundamental types. A more elaborate arrangement is shown in Figure 7.13A. Here we have a total of 12 gates in a series-parallel combination. Gates *A, B,* and *C* are in series, but this series subcircuit is in parallel with series connected gates *D, E,* and *F*. In turn, these gates form a series subcircuit in parallel with series switches *G, H,* and *I*, and they are shunted across series switches *J, K,* and *L*.

An electrical impulse traveling toward the gates from point *A* has four possible paths to follow. It can move through gates *A, B,* and *C* assuming these are in their 1 or true condition. We can write the Boolean expression as:

$$(A \wedge B \wedge C) \vee (D \wedge E \wedge F) \vee (G \wedge H \wedge I) \vee (J \wedge K \wedge L)$$

In an extensive switching circuit such as this one, it is quite possible that there will be a number of identical gates and so we can rearrange them to take on the appearance of the circuit in Figure 7.13B. No change has been made in the switching setup, but the four identical switches are now all labeled with the letter A. We can write the Boolean statement for Figure 7.13B as:

$$(A \wedge B \wedge C) \vee (A \wedge E \wedge F) \vee (A \wedge H \wedge I) \vee (A \wedge K \wedge L)$$

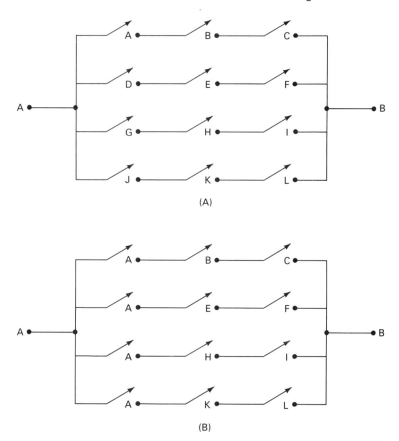

(A)

(B)

Figure 7.13. Series-parallel gating circuit (A); modification of the circuit (B).

Since the letter A is now common to each of the individual subcircuits, we can rewrite the algebraic expression as:

$$A \wedge (B \wedge C) \vee (E \wedge F) \vee (H \wedge I) \vee (K \wedge L)$$

The interpretation of this expression is that an electrical impulse can move from point A to point B by passing through switches A AND B AND C, OR through switches A AND E AND F, OR through switches A AND H AND I OR through A AND K AND L. We can simplify the original Boolean statement by taking these changes into consideration and writing the equivalent Boolean statement.

$$(A \wedge B \wedge C) \vee (A \wedge E \wedge F) \vee (A \wedge H \wedge I) \vee (A \wedge K \wedge L)$$

$$\equiv A \wedge (B \wedge C) \vee (E \wedge F) \vee (H \wedge I) \vee (K \wedge L)$$

Because the two Boolean statements are equivalent, both can be read in the same way.

As indicated earlier, the components contained in a logic circuit may all be incorporated into an integrated circuit or the gates may be used as separate parts. Information concerning the IC may be supplied in the form of a circuit diagram or as a Boolean statement using AND-OR symbols. If it is a Boolean statement it should be possible to reconstruct the basic diagram using SPST switch symbols.

INTERPRETING BOOLEAN AND-OR STATEMENTS

The symbol ∧ or the word AND always means a series circuit. A statement such as $A \wedge B \wedge C$ means we have three switches in series, three because we have three letters identifying the switches, and series because the AND symbol is used to associate the letters.

The symbol ∨ or the word OR always means a parallel circuit. A statement such as $A \vee B$ means we have two switches, two because we have two identifying letters, and parallel because we have used the OR symbol, V. OR and AND symbols can be used in a variety of combinations to supply series-parallel circuits.

As an example, consider the Boolean statement:

$$A \wedge (B \wedge C) \vee (E \wedge F) \vee (H \wedge I) \vee (K \wedge L) = B$$

To interpret this statement, follow these steps:

1. There are four terms in parentheses. This means there are at least four subcircuits.
2. There is one term not in parentheses. This single term, the letter A, indicates the presence of another subcircuit. There is now a total of five subcircuits.
3. Examine the number of letters in each subcircuit term. In the example given there are two for each subcircuit. This means that each subcircuit contains two gates.
4. Examine the symbol connecting the letters inside the parentheses. In the example given the symbol is ∧. This AND symbol means the subcircuit is a series circuit. Since the same symbol is used in each subcircuit, they are all series circuits.

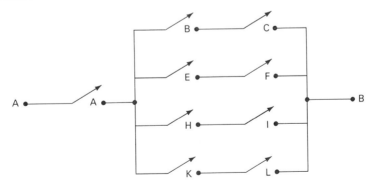

Figure 7.14. Circuit derived from Boolean statement.

5. Finally, examine the letter not enclosed in parentheses. It is a single letter and so represents a single gate. It is joined to the other subcircuits by the AND symbol, ∧, and so is in series with them.

From the information supplied it is now possible to draw the simplified gating circuit (Figure 7.14).

ADVANTAGES OF BOOLEAN ALGEBRA

Boolean algebra applied to gating circuits has a number of advantages. It means the gating circuit need not be drawn or even constructed initially. All the work can be done, and then the final circuit can be obtained from the results of simplifying the original Boolean statement. The circuit can then be drawn, using symbols for mechanical SPST switches. For very complex gating circuits it is often convenient to avoid drawing the circuit and to concentrate on obtaining the simplified expression.

Boolean algebra does not mean the equivalent electronic circuit can be drawn at once from the simplified circuit. There are other factors to be considered, including the type(s) of transistors to be used, phase inversion (if required), diode clamping (if required), amount of bias, amount of signal amplification needed, and so on.

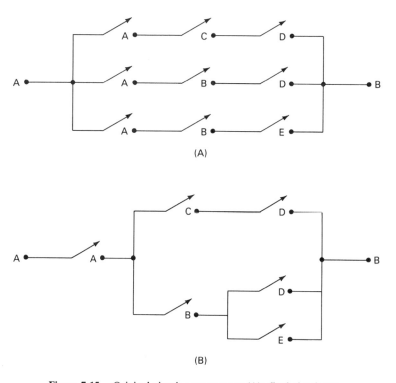

Figure 7.15. Original circuit arrangement (A); final circuit (B).

But these are factors that can be worked on after the simplified switching circuit is obtained.

Figure 7.15A shows the start of gating circuitry. Nine switches are required, but these can be reduced to six by sorting the gates into those that are identical, and then using Boolean algebra to arrive at a final circuit determination (Figure 7.15B).

PULSE POLARITY

Assuming a pulse rests on a line representing zero voltage, a pulse moving upward from this line can be said to be moving in a positive direction, or is a positive pulse. Similarly, a pulse moving downward does so in a negative direction and is a negative pulse. Since the input to the gates is a pulse, what we have in gating circuits are several facts: the first is whether the input will or will not result in an output; the polarity of the output; and the phase of that output, that is, whether the output will be in phase (move in the same direction) or out of phase (move in an opposite direction) compared to the input.

THE NAND GATE

A NAND gate, symbolized in Figure 7.16A, is a negative AND gate, also known as a NOT AND circuit, or it could be considered as an AND gate followed by an inverter. Its truth table (Figure 7.16B) is exactly the opposite of that used for the AND gate. NAND is a contraction of NOT AND. The NAND symbol is an AND symbol with a small circle to indicate its negative function.

A	B	OUT
0	0	1
0	1	1
1	0	1
1	1	0

(A)   (B)

Figure 7.16. NAND gate symbol (A); truth table (B).

Figure 7.17 is an illustrative circuit for demonstrating the NAND function. An electrical pulse arriving at point A will move through the lamp if one or more of the switches *A*, *B*, and *C* are open. Without the lamp the circuit would be an AND type in which switches *A* AND *B* AND *C* would need to be closed for a pulse to

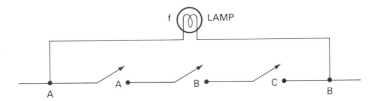

Figure 7.17. Circuit representing the NAND function.

arrive at point B. With the shunting lamp in the circuit, a pulse will arrive at B only
if all the switches are open. While the symbol for the NAND gate in Figure 7.16 shows
only two inputs, it is possible to have three (or more). The corresponding symbol
would have three inputs, which could be identified as *A*, *B*, and *C*, and an output
terminal having any other letter. However, the lower-case letter "f" is often used.

The NAND circuit can consist of three transistors in series (Figure 7.18) and
in this case they are P-N-P types. These transistors, connected in series, conduct when

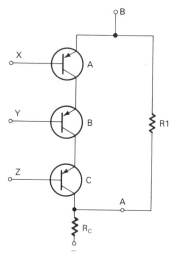

Figure 7.18. NAND circuit using
bipolar transistors.

the collector and the base are negative with respect to the emitter. R1 is the load resistor and is shunted across the transistors. Each transistor can receive input pulses
at points marked X, Y, and Z. Depending on the polarity and strength of these pulses,
one or more of the transistors can be made to conduct. As long as any one of the
transistors is cut off, an input signal at point A will travel through R1.

In still another arrangement (Figure 7.19), an AND gate is followed by an inverter. This particular gate has two inputs, but other circuits could have three or more.

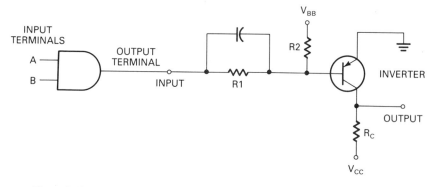

Figure 7.19. AND gate followed by inverter circuit supplies NAND function.

If none of the transistors in the AND gate permit the passage of a signal, the voltage at the top of the load resistor, R_c, rises to a maximum value. If the gates are conductive, that voltage drops to a very low amount.

The polarity of the output pulse in the AND circuit of Figure 7.19 is inverted by having the pulse fed into a grounded-emitter amplifier. The behavior of this circuit is shown in more detail in Figure 7.20 except that an N-P-N transistor has been substituted for the P-N-P. For the N-P-N to conduct means it must be supplied with a positive voltage on its base. In the absence of such a voltage the transistor is cut off and is in its 0 state. Under this condition the voltage at point B is the same as the supply voltage. With the transistor cut off there is no current flow through R2, hence no voltage drop across it.

There are three voltage possibilities at input point A, other than a peak positive voltage. The voltage could be zero, slightly positive but not enough to drive the transistor into conduction, or negative. In all three instances the input could be referred to as 0, or false, or low. The voltage at point B could be termed as high, or 1, or true. The voltage conditions at the input and output are opposites.

If an adequately high plus voltage pulse is applied to point A, the transistor will conduct. We will now have a voltage drop across R2, and so the voltage at point B will decrease to some small value, a fractional amount of the voltage supply. It can now be characterized as 1, or high, or true. This is the opposite of the input, which is now 0, or low, or false. The pulse waveforms at the input and output show the inverting action of this transistor. Unlike a transistor working as a linear amplifier, that in Figure 7.20 functions at two points: cutoff and saturation.

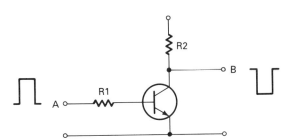

Figure 7.20. Inverter circuit.

NEGATIVE SYMBOLISM

As in the case of the AND and OR circuits, the NAND gate can also be represented by Boolean algebra symbols with a dash above a letter representing a NOT condition. For the NAND gate:

$$\overline{A} \wedge \overline{B} \wedge \overline{C} = f$$

Sometimes the equation is transposed and is written as:

$$f = \overline{A} \wedge \overline{B} \wedge \overline{C}$$

Both equations are identical. Read this equation as: f = NOT A AND NOT B AND NOT C. Sometimes, instead of using individual dashes, the equation uses just a single dash for all the inputs, and the equation is written as:

$$f = \overline{A \wedge B \wedge C}$$

All that the "not" expressions mean is that these inputs are exactly the opposite of those used for an AND circuit. Thus, if the AND circuit uses positive input pulses to obtain output at f, the NAND circuit uses negative input pulses, also to get output at f. A, B, and C represent three inputs, f is the output.

NOR GATE

Just as a NAND circuit is the inverse of the AND, so too is the NOR gate the inverse of the OR. Figure 7.21A shows the symbol and is the same as that for the OR gate but has a small circle at its end to indicate signal inversion. The NOR truth table for two inputs (Figure 7.21B) is exactly the opposite of that for the OR. The inputs for both are identical true or false statements; it is the outputs that are opposite. While the symbol shows only two inputs there can be more. Figure 7.21C shows the truth table for a three-input NOR. Using three inputs supplies greater circuit versatility. Thus, the two-input NOR has four input operating possibilities; the three-input NOR has eight. A NOR circuit can be made into an OR by following it with an inverter.

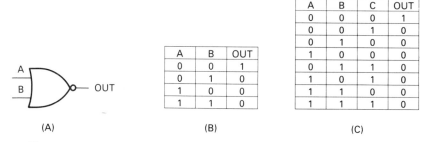

A	B	C	OUT
0	0	0	1
0	0	1	0
0	1	0	0
1	0	0	0
0	1	1	0
1	0	1	0
1	1	0	0
1	1	1	0

A	B	OUT
0	0	1
0	1	0
1	0	0
1	1	0

(A)   (B)   (C)

Figure 7.21.   Symbol for two-input NOR gate (A); truth table for two inputs (B) and for three inputs (C). A NOR gate is equivalent to a negative OR gate, sometimes called a NOT OR gate.

LOGIC CIRCUIT VARIATIONS

The AND, OR gates are fundamental logic circuits, and the NAND and NOR can be viewed as variations. However, these are not the only two possible. There are a number of others based on signal inversion at any of the inputs and/or outputs (Figure 7.22).

In Figure 7.22A the AND circuit can be modified to an OR by inverting both inputs and the output. Since both of these gates produce the same results they have

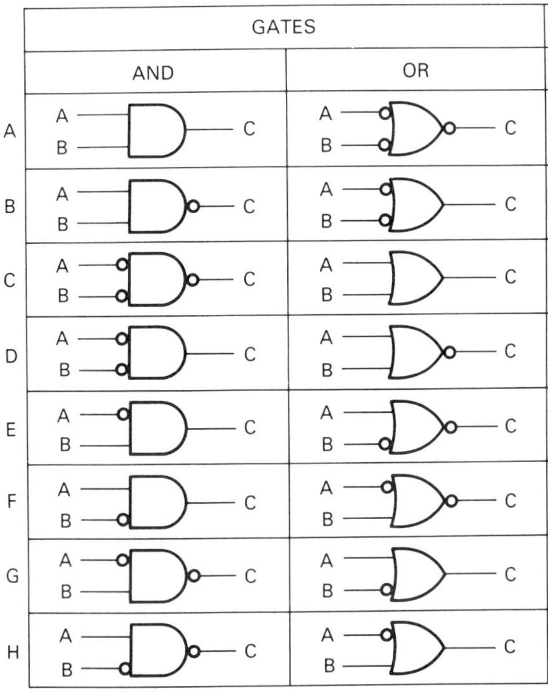

	GATES		TRUTH TABLES		
	AND	OR	A	B	C
A			0	0	0
			0	1	0
			1	0	0
			1	1	1
B			0	0	1
			0	1	1
			1	0	1
			1	1	0
C			0	0	0
			0	1	1
			1	0	1
			1	1	1
D			0	0	1
			0	1	0
			1	0	0
			1	1	0
E			0	0	0
			0	1	1
			1	0	1
			1	1	0
F			0	0	0
			0	1	0
			1	0	1
			1	1	0
G			0	0	1
			0	1	1
			1	0	1
			1	1	1
H			0	0	1
			0	1	1
			1	0	0
			1	1	1

Figure 7.22. Various AND/OR gates and their corresponding truth tables. An AND can be converted into an equivalent OR or vice versa by inverting an input (or inputs) and the output. The small circle indicates signal inversion.

the same truth table. The advantage is that it permits circuit modification, from a series arrangement for the AND to a parallel for the OR.

Figure 7.22B is a NAND but can be turned into an OR by inverting both inputs. However, its OR equivalent is not a NOR but is simply a negative input OR.

DEFINING THE OUTPUT

While any letter can be used to represent the output of any gate, that output is sometimes expressed in terms of the input. The advantage is that the output then emphasizes the type of gate being used. Thus, the output of a NOR gate can be a letter such as f, but that letter supplies no clue as to the structure of the gate. If we have a symbol, such as that shown in Figure 7.23A, and do not recognize the symbol, then there could be a need to search for further information. However, if the identifying letter or letters were supplied (Figure 7.23B), the symbol, an AND in this case, would be recognized. This technique can be applied to the gates that have been described. The truth tables can also be modified so that the output is described in terms of the input. Sometimes the truth table is supplied without an accompanying symbol. In this case the type of symbol that could have been used is easily deduced with-

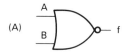

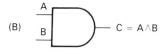

A	B	A∧B
0	0	0
0	1	0
1	0	0
1	1	1

AND GATE

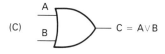

A	B	A∨B
0	0	0
0	1	1
1	0	1
1	1	1

OR GATE

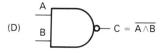

A	B	$\overline{A \wedge B}$
0	0	1
0	1	1
1	0	1
1	1	0

NAND GATE

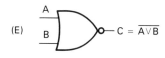

A	B	$\overline{A \vee B}$
0	0	1
0	1	0
1	0	0
1	1	0

NOR GATE

Figure 7.23.   The output in drawings B through E is supplied in terms of the inputs.

out the need for going through the results indicated in the table. Thus, in the first truth table the output at the top of the table is indicated as A ∧ B instead of using an output letter such as f. This immediately indicates that the truth table is for an AND circuit.

THE NOT NOT CONCEPT

The inversion of a signal, a positive going pulse when the input pulse is negative going, is indicated in two ways, symbolically and by the use of letters. In the symbol, a small circle at its end, either the left or the right, and attached to the symbol means signal inversion. In terms of letters, if A represents a positive going signal, then $\overline{A}$ (not A) is negative going.

A gate may consist not of one, but of a number of transistor circuits, arranged so that one follows the other. Thus, the first circuit may invert the signal, the second will invert it again, and the third will also invert, and so on. However, no matter how many stages are used, the output of the final stage can always be compared with the input of the first. If there is phase inversion when this comparison is made, then the output can be represented by a single letter with a dash line above it. If A is the input then $\overline{A}$ (not A) would be the output.

It is also possible to show that a number of inversions have taken place, and if the output is not phase inverted, the corresponding statement could be $\overline{\overline{A}} = A$. The double dash above the letter A is read as "not not" A. The two "nots" cancel; that is, a double negative is equivalent to a positive statement.

CIRCUIT DERIVATION

An integrated circuit may consist of a number of gates consisting of a variety of types: AND, OR, and so on. The objective is always to use the least number of gates that will achieve a specific objective. The approach is to start with a Boolean expression that will produce the wanted output and then to work with this expression so as to reduce it to its simplest terms.

As an example, consider a Boolean statement such as

$$f = (A \land C) \lor (A \land D) \lor (B \land C) \lor (B \land D)$$

If we were to go no further than this expression we could produce the gating diagram in Figure 7.24. Note that the output of each gate isn't represented by a single letter, which in itself conveys no information, but rather by Boolean expressions.

Each of the expressions is a gate. The first step is to look for a common letter in each. The first two gates have the letter A in common; the next two the letter B in common.

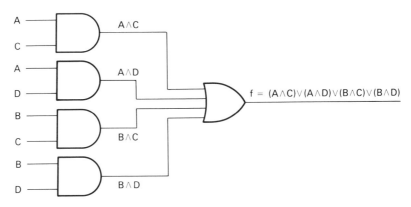

Figure 7.24. Symbolic circuit with outputs in terms of inputs.

The output can be simplified to:

$$f = A \wedge (C \vee D) \vee B \wedge (C \vee D)$$

Inside the parentheses we now have a pair of identical terms. Consequently, the equation can be further simplified to read:

$$f = (A \vee B) \wedge (C \vee D)$$

The two terms on the right-hand side of the equation mean we have reduced the original requirement of four gates to two. Each of these two gates is an OR, while the four original gates were AND, that is, A AND C, A AND D, and so on. However, the original statement requires that we combine these two gates, something we can do with the use of an AND gate. Figure 7.25 shows the final gating arrangement.

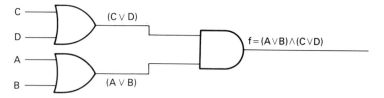

Figure 7.25.   Final Boolean arrangement of the symbolic circuit in Figure 7.24.

LOGIC CIRCUIT DEVELOPMENT

Logic circuits have gone through an electronic evolutionary process, and the factor that caused the change is time. A gate is a switching circuit and so requires time to open and close, and while the operating time is measured in nanoseconds, it is helpful to have gates work as close to instantaneously as possible. A nanosecond is one billionth of a second, or, in terms of exponents 10^{-9} second. The abbreviation is nsec. Light travels at approximately 186,000 miles a second or a distance of 1 foot in 1 nsec.

RESISTOR-TRANSISTOR GATES

Transistors used in gating circuits may seem to resemble amplifiers, but their working conditions are quite different. Such transistors work at extremes: either at cutoff, a condition in which there is no current flow through the transistor, or at saturation, a condition of maximum current. In neither of these states is there amplification.

The first gate to achieve wide acceptance but now rarely used is the resistor-transistor gate, commonly known as resistor transistor logic (RTL). It is characterized by the fact that a resistor is used in the base input of each transistor, shown as R1 and R2 in Figure 7.26. The input signal passes through each of these resistors, driving the transistors into maximum conduction. As a result there is a substantial voltage drop across the single load resistor, R3. Because of this drop, the voltage on the collectors (wired together) drops to some low value. In the absence of an input signal, the IR drop across R3 becomes much less, and so the collector voltage

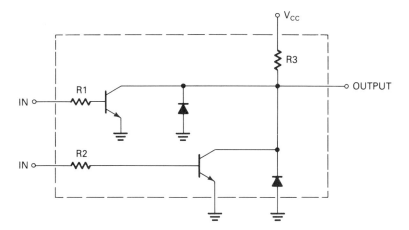

Figure 7.26.   Resistor-transistor logic (RTL) circuit.

rises. The output, then, is either a high voltage or a small one, equivalent to a pulse. These are the only two working conditions.

Transistors have internal capacitance, previously described as phantom capacitance, and this capacitance exists between the collector and emitter. This capacitance takes priority over any external components connected to the transistor and must be charged on application of an input pulse driving the transistor to saturation, and it must be discharged when the transistor is driven to cutoff. This internal capacitance works together with R1 and R2, forming an R-C network, and so the time of charge and discharge depends on the values of R and C. Thus, the operating time of the RTL logic gate becomes slower because of the presence of R1 and R2, and typically the RTL has a switching speed of 50 nanoseconds. The gating network in Figure 7.26 is no longer used but you may find it in some older circuit diagrams.

DIODE-TRANSISTOR GATES

Also known as DTL (diode-transistor logic) this gating circuit (Figure 7.27) substitutes diodes for the resistors used in RTL. The input diodes, D1 and D2, have a very low forward resistance and so the switching action of the gates can be more rapid. Typical switching time is about half that of the RTL, that is, in the order of approximately 25 nsec.

The output, C, is taken from the collector of the transistor, and so, as indicated by the upper of the two logic symbols, is inverted with reference to the input. The upper symbol is that of a NAND gate. It can be replaced by a NOR gate by inverting both inputs. The Boolean statement for the upper diagram is: A AND $B = \overline{C}$ and it can be stated as: A AND B equals NOT C. For the lower diagram we have: $\overline{A}$ OR $\overline{B} = C$. This was previously indicated in Figure 7.22B. The dashed line above the two letters for the inputs of the OR gate are opposite those of the inputs for the

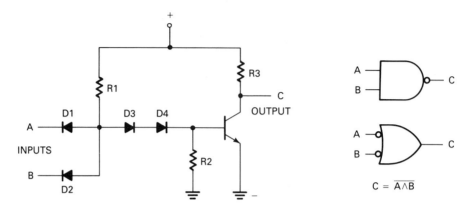

Figure 7.27. Diode-transistor logic (DTL) circuit.

AND gate. Each of the input letters uses a dash to indicate a NOT condition. Like the RTL the DTL may still be found, but usually in older circuitry.

TRANSISTOR-TRANSISTOR GATES

Also known as transistor-transistor logic, or TTL, this type of gating arrangement has replaced RTL and DTL. The gate uses a transistor that has multiemitter elements (Figure 7.28). Despite this construction the transistor is still a member of the bipolar transistor family and in this illustration is an N-P-N type. The TTL is sometimes referred to as T^2L or T-squared L.

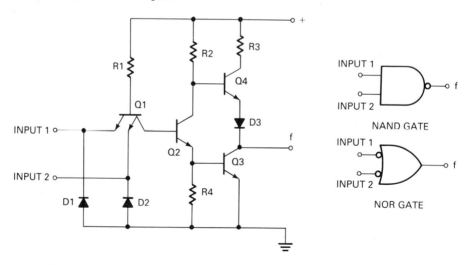

Figure 7.28. Transistor-transistor logic (TTL) circuit. Depending on polarity of input and output it can be a NAND or NOR, as indicated by the symbols.

Adjacent to the circuit are two logic symbols. The output transistors form an AND gate (upper symbol) and since it has an output that is out of phase with its input, is a NAND. It can also work as a NOR gate by inverting both pulse inputs. The resulting symbol is at the lower right. When used as a NAND the Boolean expression is a A AND $B = \bar{f}$. The dashed line above the letter f indicates that the output is the inverse of the input.

When used as a NOR the Boolean equation is: $\bar{A}$ OR $\bar{B} = f$. Both inputs are inverted. Cutoff in the transistor does not reach zero, and as a consequence this reduces the amount of time needed to reach saturation. The switching speed is much lower than a DTL and is often less than 10 nsec.

EMITTER-COUPLED GATE

This gate, possibly better known as emitter-coupled logic (ECL) is somewhat similar to a differential amplifier (Figure 7.29). It is even faster in operation than the TTL but its operating power requirements are higher than any of the logic gates that have been described and so it may be necessary to mount the IC on a heat sink.

The transistors do not reach saturation. Transistor Q3 has a reference voltage established on its base. This voltage determines the limitations of the input signal.

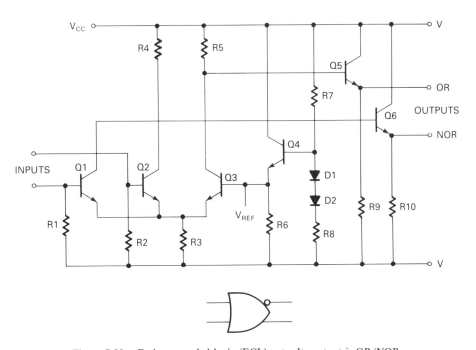

Figure 7.29. Emitter-coupled logic (ECL) gate. Its output is OR/NOR.

The symbol for the ECL gate is a modified version of the OR. Unlike that symbol this one shows two outputs. One of these is OR and the other is negative OR, that is, NOR.

FAN-IN AND FAN-OUT

For the sake of simplicity in drawing a circuit diagram, the number of inputs or outputs of a gate is limited and is seldom shown as more than three. However, a larger number of inputs can be used. Thus, the utility of a gate will depend partially on the number of inputs the gate can accept and still function correctly. The quantity of inputs is referred to as fan-in (Figure 7.30). If, for example, a gate could safely accommodate just 16 inputs and a circuit called for 32 inputs, a pair of OR gates could be connected as indicated in the figure. Each of the gates having a fan-in of 16 would have a single output. A pair of such gates would then have two outputs, which could be input to another OR gate.

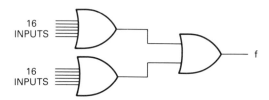

Figure 7.30.   Fan-in technique.

A fan-out has a requiremment that is opposite that of a fan-in. It describes the number of circuits that can be supplied with input signals from a gate. Field-effect transistors are suitable for fan-out use since they have high-input impedances and consequently require very little power.

Multiple Emitter Logic

This type of logic circuit uses a transistor having two or more emitters, with the one shown in Figure 7.31 equipped with four. In this circuit these transistor elements are used to supply four inputs identified as *A, B, C,* and *D* with the circuit arranged as a NAND gate using bipolar transistors. The output is:

$$f = \overline{A \wedge B \wedge C \wedge D}$$

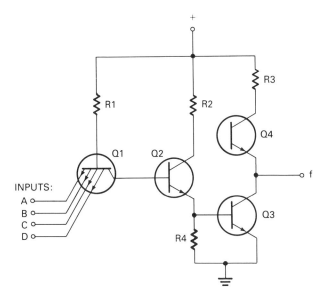

Figure 7.31. Multiple emitter input logic circuit.

Waveforms and Miscellaneous Circuits

A circuit diagram may consist of a number of symbols representing various parts interconnected by lines but not supplemented by additional information. At the other extreme there are diagrams that are extensively detailed with notes, with identifying data, parts lists, and waveforms.

The waveforms may appear at the signal input of a circuit, at the signal output, or both. They may be drawings or photos, or a combination of the two. In some instances the waveforms also include data such as peak-to-peak voltage. Waveforms, with or without supplementary information, can be regarded as an integral part of circuit diagrams.

CIRCUIT DIAGRAMS

There is no precise definition of a circuit diagram that is all-inclusive, that covers fundamental circuits, subcircuits, partial and complete circuits. It is possibly for this reason that no census has ever been taken of the total number of circuits, but as a guess there must be hundreds or more.

Because there are so many, it would be impossible to describe all of them and in some instances it would be difficult even to categorize them. However, illustrating a number of miscellaneous circuits, plus a description of the more common waveforms, will be helpful in developing greater familiarity with the symbols and also with the numerous possible arrangements of subcircuits and partial circuits.

WAVEFORMS

A variety of waveforms are used in electronics, particularly in circuits involving television receivers. The use of waveforms is a considerable aid in servicing when an oscilloscope is available.

Sine Wave

The sine wave is the basic waveform from which the others are derived. There are various ways of defining this wave: as a periodic trigonometric curve representing $y = \sin X$, with values repeated every $360°$; as a wave that can be expressed as the sine of a linear function of time; as an AC wave having a peak value twice during each complete cycle. The development of a sine wave from an electromechanical generator is pictured in Figure 9.28 in the following chapter.

The RMS (root mean square value) of a sine wave (Figure 8.1A) is 70.7 percent of either peak; the average value is 63.6 percent of the peak. The basic sine wave is sometimes referred to as a fundamental wave or first harmonic. Each higher order harmonic is also a sine wave and can consist of a second harmonic (fundamental sine wave frequency multiplied by two); third harmonic (fundamental frequency multiplied by three) and so on. Combinations of a fundamental plus several or more harmonics (Figure 8.1B) result in waves of various shapes, not sinusoidal, but which can be resolved into the fundamental sine wave and its harmonics.

The time duration of one complete cycle is the period of the wave and is inversely proportional to its frequency. In terms of a formula, $t = 1/f$; f is the frequency of the wave in Hertz, and t is the time in seconds. The wavelength, inversely proportional to the frequency is the distance measured from one peak to the adjoining peak of the following cycle.

Sine waves can be generated electronically or mechanically with typical electronic circuits including the Hartley, Colpitts, crystal-controlled, Armstrong, and so on. Line voltage in the home, produced electromechanically by an AC generator, is a sine wave. Radio-frequency carriers for AM, FM, or TV signals are electronically produced sine waves.

Clocks

A clock is a reference timing source supplying pulses that can be used as synchronizing triggers for circuits whose operation must be time-controlled. The operation of a clock is expressed in terms of frequency, i.e., mHz, and is known as its clock rate.

Timing can be obtained from an oscillator whose frequency can be crystal-controlled for optimum accuracy. Frequency dividers are used to obtain highly accurate values.

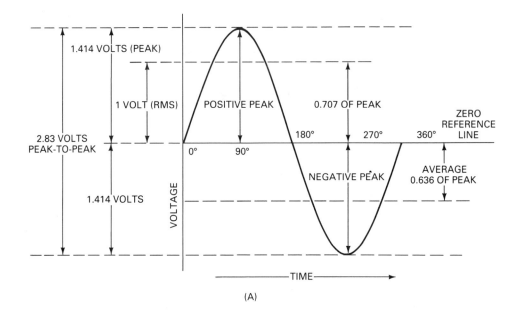

(A)

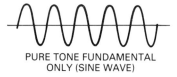

PURE TONE FUNDAMENTAL
ONLY (SINE WAVE)

FUNDAMENTAL PLUS SECOND
HARMONIC

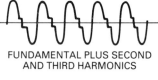

FUNDAMENTAL PLUS SECOND
AND THIRD HARMONICS

(B)

Figure 8.1. Sine wave voltage or current (A); sine wave plus second and third harmonics (B).

Pulse Waveforms

These are sometimes referred to as square waves, but more often are rectangularly shaped. In a square wave the voltage or current rises extremely sharply (theoretically instantaneously) to some predetermined maximum value and holds this value without variation (Figure 8.2). The rise time of the wave is identified by the leading edge of the waveform (Figure 8.2A). The trailing edge indicates the rapid fall of the pulse to zero. The time duration (Figure 8.2B) of the pulse (t_d) is expressed in some unit of time such as a microsecond, millisecond, or second. One complete cycle of a pulse wave is the time from the start or leading edge of one pulse to the start or leading edge of the following pulse.

The word "frequency" found in connection with sine waves is generally not used for pulses. Instead, the number of cycles per unit of time is the pulse repetition

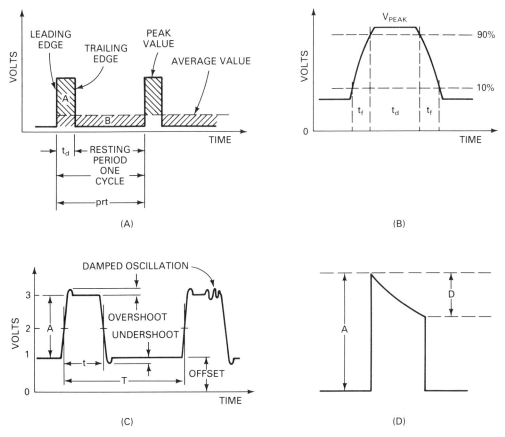

Figure 8.2. Pulse waveforms. Pulse characteristics (A); rise and fall times (B); overshoot and undershoot (C); percentage of droop (D).

rate (prr) and is the number of complete pulses per second. The time required for the completion of a cycle is the pulse repetition time or prt.

There is a difference between (t_d), the time duration of a pulse and prt, the pulse repetition time. The time duration of a pulse is the time required for the pulse to rise, remain, and fall. Pulse repetition time includes t_d plus the resting period, or the time required until the start of the next cycle.

Like a sine wave, a pulse wave has a peak and an average value with the peak value the maximum amplitude of the pulse. The average value is obtained when the area enclosed by the pulse is equal to the imaginary area indicated by dashed lines in Figure 8.2B. The ratio between the average and peak values is the duty cycle. The duty cycle can be expressed as: duty cycle = average value/peak value. It can also be given as: duty cycle = t_d/prt = t_d × prr.

Ideally, a pulse will reach its peak instantaneously, or in zero time, and will also decrease to its minimum or zero value instantaneously, an impractical demand. The rise time of a pulse is considered as the time required for the pulse to change its voltage amplitude from 10 percent of its value to 90 percent. Similarly, fall time is the time needed for the pulse to decrease from its 90 percent point to 10 percent.

The pulse period is the time required between a pair of adjacent pulses and can be measured from one leading pulse edge to the leading pulse edge of the following pulse, or between a pair of adjacent trailing edges.

Pulse width is the time measured between the 50 percent amplitude points of any single selected pulse. The amplitude of the pulse, shown by the letter A in Figure 8.2C, is the strength of the pulse measured from its baseline to its maximum amplitude, but does not include any possible undershoot or overshoot of the waveform.

A pulse can start at 0 volt and rise to some predetermined amplitude, followed by a sequence of similar pulses. However, it is also possible for the initial strength of the pulses to be at some value other than zero. The voltage difference between zero and that initial value is known as pulse offset and can be positive or negative. Positive polarity is assumed unless otherwise indicated.

Ideally, a pulse can reach its maximum and minimum voltage values without exceeding them. The extent to which it does so is known as overshoot and undershoot (Figure 8.2C). The amount of overshoot or undershoot can be expressed as a percentage of the pulse voltage amplitude. Small percentages are desirable.

Pulses can be positive or negative and are independent of the amount of offset voltage. Positive pulses are usually drawn above the voltage baseline; negative pulses below. Negative pulses can also be shown in the same form as positive pulses; that is, above the voltage baseline provided the polarity is indicated in some way.

Pulses in an inductive circuit can produce an effect known as ringing, a damped oscillation at the end of either the leading or trailing edge, or both.

Pulses are used as part of the composite video signal and include vertical and horizontal synchronizing pulses, blanking and equalizing pulses. They are also used in logic circuits (Chapter 7).

The droop of a pulse is its loss in amplitude from its leading edge to its trailing edge. The percentage droop is the ratio of the amplitude of the trailing edge to that of the leading edge (Figure 8.2D). In terms of formula: % Droop = D/A × 100.

Sawtooth Waveforms

A sawtooth waveform, (Figure 8.3) can be generated through the charge and discharge of a capacitor. Each part of the wave is called a "ramp." The trace ramp occurs during the time the sawtooth rises from some minimum to maximum value; the retrace ramp during the time the sawtooth decreases from maximum to minimum. Ideally, the retrace ramp should occur in zero time, but some time is required. In a sawtooth the trace ramp, sometimes called the positive ramp, and the retrace ramp, sometimes called the negative ramp, do not have equal slopes, with rise time greater than fall time.

A sawtooth voltage is the type applied to the horizontal deflection plates of an oscilloscope. It is also used in the development of a suitable waveform for the vertical and horizontal deflection coils in the yoke around the neck of a TV picture tube.

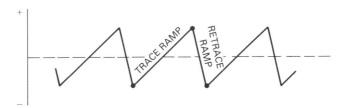

Figure 8.3.   Sawtooth waveform.

Ramp Wave

The ideal sawtooth waveform rises linearly from zero until it reaches a peak, at which time its drop to zero takes place almost at once. However, as in other waves, there can be distortion, possibly caused by a nonlinear rise. Thus the rise may belly out. The drop to zero may take measurable time. A wave that does not have these characteristics is a practically perfect sawtooth having a linear rise and a drop to zero that is close to perpendicular. This perfect sawtooth is called a "ramp wave."

Trapezoidal Waveforms

The trapezoid is a combination pulse and sawtooth wave (Figure 8.4A). When a sawtooth current is sent through an inductor (L) in series with a resistor (R), the sawtooth current will produce a sawtooth voltage (Figure 8.4B) across the resistor. However, the voltage appearing across the coil will have a pulse shape. When these two

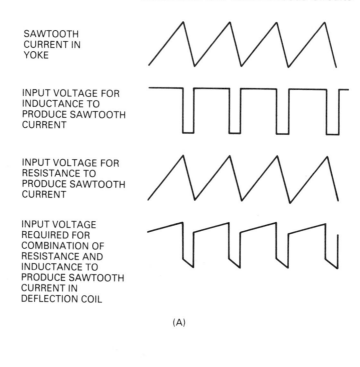

(A)

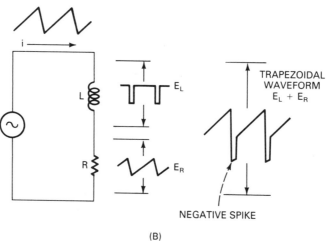

(B)

Figure 8.4. Steps in the formation of a trapezoid.

are combined $(E_L + E_R)$ the resultant wave is a trapezoid and is used across the deflection yoke of a picture tube (B). The trapezoidal current produces a magnetic field which increases linearly to some maximum and then drops to zero, producing linear sweep of the cathode-ray beam across the face of the picture tube. Figure 8.5 shows the appearance of the sawtooth current as it appears in the deflection coils. The sawtooth retrace takes place during the time blanking pulses are applied to the tube. As a result, retrace lines do not appear.

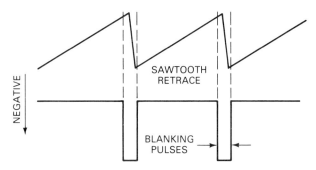

Figure 8.5. Sawtooth retrace takes place during blanking pulse time.

Stairstep Wave

Also known as a staircase waveform (Figure 8.6), the stairstep wave is used at the input of a video cassette recorder as a test signal to determine the ability of the VCR to reproduce monochrome shades ranging from black to white. It is also used to check the linearity of IF and video amplifiers as well as the alignment of synchronous detectors.

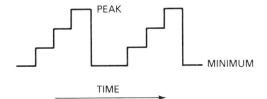

Figure 8.6. Stairstep wave.

Integrated Waveforms

An integrating circuit (Figure 8.7A) is used to separate serrated vertical sync pulses from horizontal sync pulses. The incoming pulses (Figure 8.7B) are in rectangular form. Figure 8.7C shows the resultant output waveform. It is used to synchronize the vertical sweep oscillator.

MISCELLANEOUS CIRCUTS

Waveforms are not only used to indicate the shape of a signal but also act as an indication of the function of a circuit. This can be very helpful if the purpose of a circuit isn't obvious.

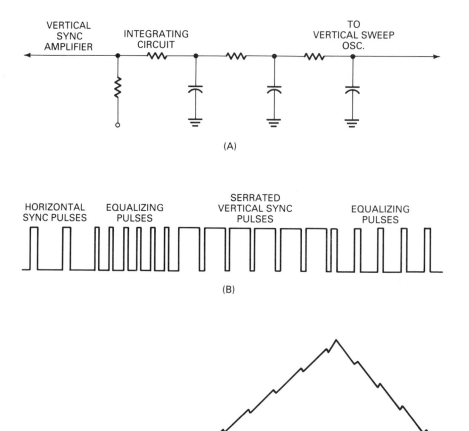

Figure 8.7.   Integrating network (A); pulse input (B); integrated waveform (C).

Noise Cancellation Circuit

The circuitry in Figure 8.8 consists of a video amplifier and a noise canceller. These circuits are identified as such in this diagram, but this is not always done. However, the use of the noise canceller can be deduced by comparing the phase of the noise pulse in its output with that of the output of the video amplifier. The noise pulses are shown as sharp spikes. They can have a varying amplitude and can be stronger or weaker than the accompanying signal.

    The video signal is delivered from the video demodulator to the video amplifer, Q1, but as shown by the waveform at its input is accompanied by noise pulses. Q1 is an emitter follower, and so the phase of its output signal is the same as that of

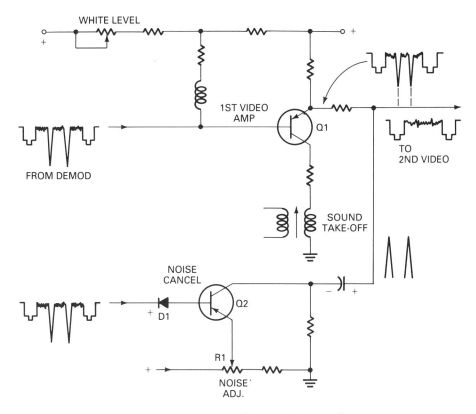

Figure 8.8. Video amplifier and noise canceller.

the input. However, the noise canceller transistor, Q2, has its output from its collector and so supplies phase reversal of its input signal.

The base of Q2 is connected to a diode, D1. This diode is biased to cutoff and so will not conduct when supplied with the video signal. However, if the noise pulses have a greater amplitude they will take the diode out of cutoff, permitting it to conduct. The output of Q2 consists only of inverted noise pulses and these are combined with the video signal at the output of Q1. Since the noise pulses are out of phase they cancel. The bias on Q2 can be adjusted by variable resistor R1, and so the functioning of Q2 can be made operative for weak or strong noise pulses.

Sync Separator

The video amplifier may consist of one or two transistors, but somewhere in that circuitry the sync pulses must be obtained from the composite video signal. Figure 8.9 shows a method of doing this. Q1 is the video amplifier; Q2 the sync seperator. This transistor amplifies the pulses and inverts them. Q2, as shown in Figure 8.10, is followed by a sync pulse limiter.

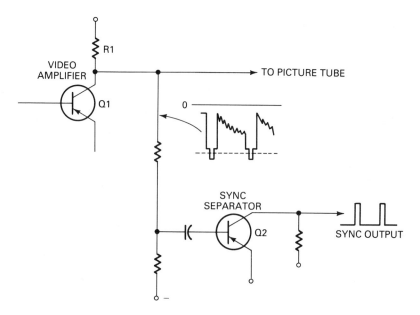

Figure 8.9. Sync separator.

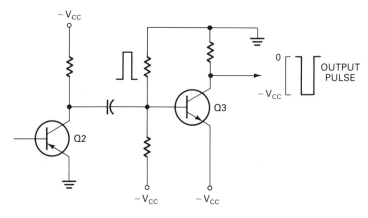

Figure 8.10. Sync pulse limiter.

Sync Pulse Limiter

Theoretically it should be possible in TV receiver circuitry following the sync separator to use the sync pulses directly. There are several reasons, however, why this isn't always practical. The sync pulses may have been accompanied by noise pulses, the pulses need amplification, and they should all have a constant amplitude. A method for doing all this is indicated in Figure 8.10. Q2 in this diagram operates between the extremes of its characteristic curve, between current cutoff and saturation. By not being limited to the linear portion of this curve, a greater output pulse amplitude is obtained. However, the input pulse signal must be able to drive Q2 between these two operating limits. One of the advantages of this type of operation is that noise pulses existing between input pulses may not have enough strength to drive Q2 and so do not appear in its output.

The first transistor in this circuit is a bipolar P-N-P and so requires a negative-going pulse to turn it on. Q2, however, inverts the signal, and so the sync pulses are positive going at its collector output. These pulses drive Q3, which supplies not only additional pulses amplification but inverts the signal.

Vertical Output

The vertical output stage (Figure 8.11) works as a single-ended class-A amplifier. The output of Q1 is an inverted sawtooth, while the voltage appearing across the yoke coils is trapezoidal. There is no phase change in the input signal since Q1 is connected as a common emitter.

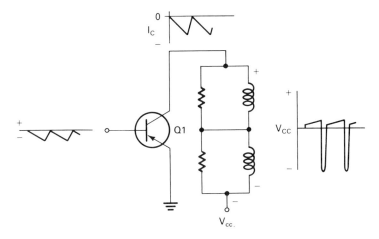

Figure 8.11. Vertical output stage.

Differentiating Circuit

An integrating network was shown in Figure 8.7 with its output used to control the operating frequency of the vertical sweep oscillator. The horizontal oscillator also requires a triggering signal, and this is supplied by a differentiating network shown in Figure 8.12A. The horizontal sync pulses (B) applied to the input of the differentiator result in sharp pulse spikes (C), which are then used to control the operating frequency of the horizontal oscillator. The horizontal pulses are the same distance apart and continue through the vertical blanking period; consequently the horizontal oscillator does not lose sync during this time.

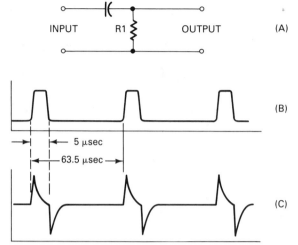

Figure 8.12. Differentiating circuit (A); waveform of horizontal sync pulses at input to differentiator (B); horizontal sync pulses following differentiation (C).

WAVEFORMS AND THE BLOCK DIAGRAM

The circuit diagram of a complete TV receiver is often quite extensive but can be considerably simplified by the use of IC symbols and blocks to represent various functions. Usually waveforms are included at various points in the block diagram (Figure 8.13). Such a diagram is helpful for learning the functioning of a TV set, for getting an overview of a particular model or as an aid in working with a scope.

This block diagram is that of the video signal stages, starting with the video signal demodulator and extending to the cathode video signal input points on the picture tube.

Clipper Circuitry

This circuit is used to remove some part of an input waveform. The circuit may be a top clipper, removing only the upper portion of a waveform, or a bottom clipper, removing only the lower part, or a combination of the two. Clippers (also called "limiters") can be diodes or transistors and both may (or may not) be biased for the work they are to do. Clippers can be used in connection with frequency-modulated signals to remove any unwanted amplitude variations.

Figure 8.14 shows diodes in various circuit arrangements. Sine wave input is used throughout but it can be any waveform that is to be rectified and whose peaks are to be trimmed. Figure 8.14A is a positive series limiter, series because diode D1 is in series between the input and output and positive since the positive half of the input is removed. The circuit arrangement is the same as that of a half-wave rectifier used in a power supply.

Figure 8.14B is similar to that of A but with the diode transposed; it is referred to as a negative series limiter. In both instances in the first two circuits there is a

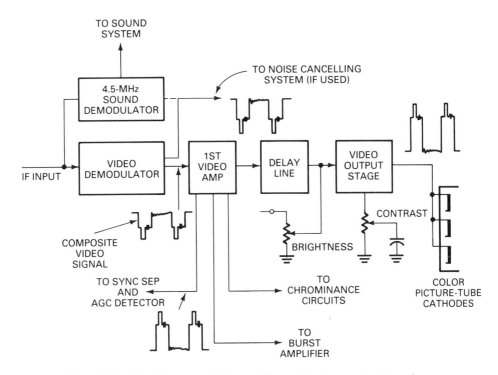

Figure 8.13. Block diagram of video amplifier section plus associated waveforms.

signal loss across the diode (E_D), and as a result the clipped waveform does not have the amplitude of the input signal. The amount of loss is indicated in the output by the dashed line across the wave and identified by E_D. Diodes can also be used as parallel limiters with the diode in shunt across the output. Figure 8.14C shows top limiting; D shows bottom limiting.

Figure 8.15 shows a pair of diodes shunted, but in reverse, and with a bias voltage applied to each. The amount of clipping is determined by the characteristics of the diodes and the amount of bias voltage applied. The horizontally dashed line in the output waveform shows the effect of top and bottom clipping.

Transistor Clipper

The arrangement in Figure 8.16 uses a P-N-P transistor that has no forward bias. As a result the only collector current is a small amount of leakage current from the collector to the emitter. The circuit operates only when the transistor is taken out of current cutoff, something that takes place when the input pulse moves in a negative direction.

At the time the transistor conducts, some current flows through resistor R1, making the top end of R1 minus and the other end plus. Another not apparent in the diagram, capacitor C1 is shunted across R1 through an external circuit (not shown)

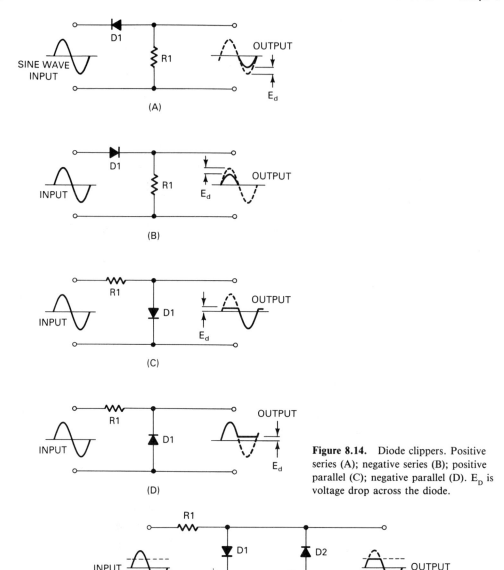

Figure 8.14. Diode clippers. Positive series (A); negative series (B); positive parallel (C); negative parallel (D). E_D is voltage drop across the diode.

Figure 8.15. Double diode clipper.

connected across the input. C1 becomes charged with its negative terminal connected to the base.

When the negative input pulse is completed, C1 discharges through R1, but the direction of current flow is down through R1. Since this resistor is connected be-

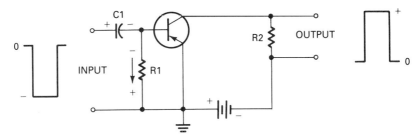

Figure 8.16.   P-N-P clipper.

tween base and emitter, a condition of forward bias exists. This forward bias is overcome by the positive portion of the input pulse. By selecting various values of C1
and R1 it becomes possible to have control over the amount of clipping to be done
by the circuit. If C1 and R1 are omitted, the circuit will clip half of the input pulse
but less than half if C1 and R1 are in the circuit.

Power Supplies

Power supplies were described in Chapter 4 but can now be considered in terms of
waveforms. The input to a power supply is sine wave AC (Figure 8.17A). For a half-
wave rectifier, current flow through the output load resistor R is in the form of one
pulse for each complete AC cycle supplied to the input. This means that for this rec-

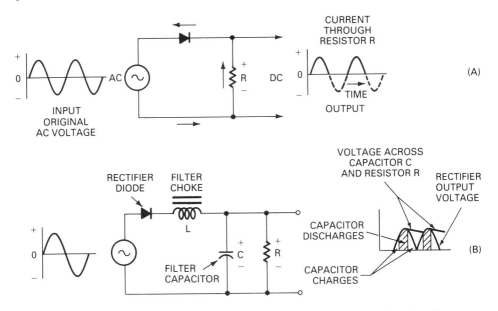

Figure 8.17.   Input and output waveforms of half-wave rectifier (A); alteration of
output waveform when filter is used (B).

tifier, assuming 60 Hz input, the output consists of 60 pulses, with a time interval between each pulse. This is the waveform that follows the rectifier but precedes the filter. The dashed line portion of the output wave indicates that part of the input waveform that has been removed.

Figure 8.17B shows the effect of a filter on the pulse waveform. Capacitor C charges during the time current flows through load resistor R. When the polarity of the AC input voltage reverses, the diode blocks the movement of current from the power line. Capacitor C discharges through resistor R. As a result the voltage output tends to become more continuous and less of a series of pulses. Choke coil L also helps smooth the output voltage.

Relay Control Circuit

Figure 8.18 shows a P-N-P transistor used as a relay control. In the absence of a pulse input signal, the transistor operates class B. Under these conditions practically no current flows through the transistor or through the relay. As a consequence there is little or no magnetic field around the relay coil, and so the relay contacts remain in their open position. With the arrival of a negative going pulse, the P-N-P transistor becomes forward-biased with current moving from the negative terminal of the B-supply through the relay coil to the collector of the transistor. The armature of the relay is attracted to its opposite contact and the relay closes. This, in turn, can close some circuit connected to the relay terminals.

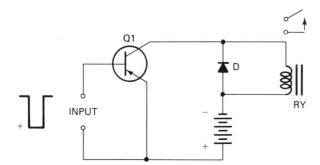

Figure 8.18.   Pulse-operated relay.

When the input pulse drops to zero, the current flowing through the relay coil decreases equally abruptly, inducing a large counter electromotive force across the relay coil. This voltage could be large enough to destroy the transistor by exceeding its inverse peak voltage rating. This is prevented by diode D shunted across the relay coil.

CHOPPERS

One of the disadvantages of nonvarying DC is that it cannot be used in connection with a transformer to be stepped up or down. A transformer doesn't necessarily require an AC input. The input can be DC provided it can vary, producing a comparable varying magnetic field around the transformer's windings.

Interrupting the DC supply can be illustrated by the circuits shown in Figure 8.19A and B. In A, a switch SW is shunted across the DC supply. When the switch is open current flow is limited by R1 and R2 in series. When closed, R2 is shorted by the switch and current flow increases. Turning the switch on and off in sequence will produce a varying direct current through R2, referred to as chopped output.

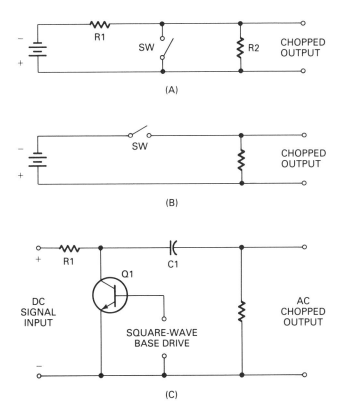

(A)

(B)

(C)

Figure 8.19. Shunt chopper (A); series chopper (B); electronic chopper (C).

A similar circuit is shown in B and has the advantage that one of the resistors has been omitted. The operation of this circuit is the same as that of A. Both, however, are for demonstration only since manual operation of the switch is impractical.

A better arrangement is the solid-state circuit in Figure 8.19C using an N-P-N transistor. A square wave pulse is applied between the emitter and base, alternately pulsing the transistor between cutoff and saturation. When the transistor is on, maximum current flows through R1, and there is a large voltage drop across it. When the transistor is cut off, there is no current through R1, and the maximum voltage appears across the load resistor. The signal appearing at the output will be a square wave whose amplitude is about that of the DC signal input.

AMPLIFIER CLASSES

While there are other amplifier classes, these circuits can be generally categorized as A, B, or C. The operating class of an amplifier is determined by the amount of bias and its polarity. With certain exceptions, amplifiers can be used for AF or RF.

Figure 8.20A illustrates a class A amplifier. This circuit uses an N-P-N and the polarity of its bias and its amount is such that current flows through the load resistor for the entire time of the input signal. It is sometimes referred to as single-ended since it operates with just one transistor. The output signal is amplified and is out of phase with the input. An amplifier of this kind can supply a quality signal, that is, one with minimum distortion, but its efficiency is low. The transistor is biased so that it operates at the center of the linear portion of its characteristic curve.

The drawing in B is that of a class B amplifier. Although just a single transistor is shown, it is more commonly operated using a pair in push-pull. The bias battery has been removed, and so this unit has no forward bias. As a result it remains cut off until the polarity of the input signal is such as to take it into conduction. When this happens, current flows through the load but is available only during half of the input cycle. Only half of the output voltage wave exists and, in effect, with single-ended class B, the transistor works as a rectifier. To work as an amplifier a pair of transistors are required, with each working on individual alternations of the input cycle. The efficiency of the class B amplifier is better than that of class A.

The class C amplifier (Figure 8.20C) is biased beyond cutoff. The transistor does not function for most of the input cycle, depending on the amount of bias and the amplitude of the input signal. The output consists of a series of disconnected signal pulses. While the efficiency of this amplifier type is better than either class A or B, it isn't suitable for audio amplification but does find application in RF amplifiers in continuous wave (CW) transmitters.

Figure 8.20D shows the reassembly of the input waveform by a push-pull amplifier operating class B. Class B amplifiers work on the assumption that both transistors in the circuit are identical. Unless they have very close characteristics, the result can be crossover distortion, as indicated in Figure 8.20E. In this condition the two halves of the output waveform do not match exactly, and the result is distortion at the crossover point. This can be countered in two ways. One is by carefully selecting transistors so their characteristics are as alike as possible. Such transistors are sold as matched pairs. The other method is to adjust the bias of each transistor while checking the output with an oscilloscope.

PULSE INVERTER

A pulse inverter (Figure 8.21) is a circuit that inverts the phase of an input pulse and at the same time supplies signal amplification. If 0 is taken as the reference, the input pulse moves in a positive direction producing an output pulse that moves negatively. If the input pulse is reversed, so is the output. The size of the output pulse is controlled by two factors: the gain of the transistor and the amount of bias sup-

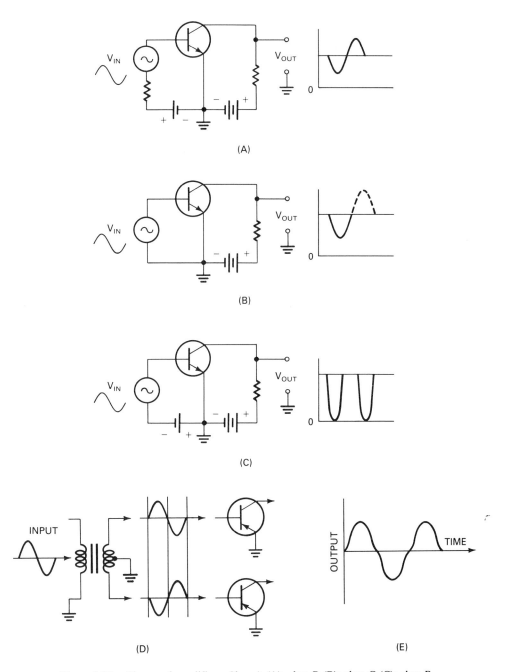

Figure 8.20. Classes of amplifiers. Class-A (A); class-B (B); class-C (C); class-B pushpull (D); crossover distortion (E).

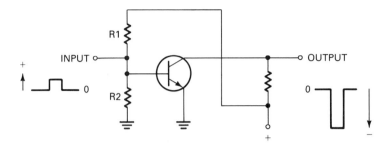

Figure 8.21.   Pulse inverter.

plied by resistors R1 and R2. The circuit contains no reactive components, such as coils or capacitors, with these having a possible effect on the output. A single power supply is used for both collector and bias voltage.

RADIO-FREQUENCY PREAMPLIFIER

The function of the radio-frequency preamplifier in Figure 8.22 is to strengthen the input signal. In this example the signal is amplitude-modulated, and except for strength, the output signal is identical with the input. The amplifier must be capable of passing the complete band of frequencies contained within the waveform. The advantages of the field-effect transistor are that it can supply high gain, requires very little operating power, and has a high input impedance. While this circuit shows a tuned input, increasing selectivity, some amplifiers have both untuned inputs and outputs. In some arrangements the preamplifier can be used as a separate unit and

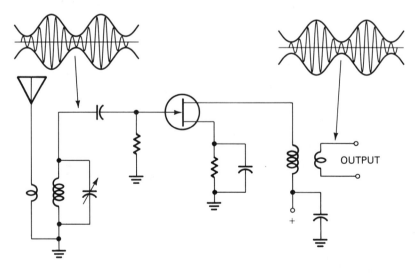

Figure 8.22.   Radio-frequency preamplifier.

is mounted on an antenna mast so as to be as close to the antenna as possible, thus improving the signal-to-noise ratio. The output can be brought into a converter, into a mixer, or into an in-receiver tuned-radio-frequency amplifier stage.

PIERCE CRYSTAL OSCILLATOR

The Pierce crystal oscillator can use either of the bipolar transistors, N-P-N or P-N-P, or, as shown in Figure 8.23, a field-effect transistor with the crystal connected between the gate and drain. It is sometimes the preferred form of crystal oscillator since it does not require a tuned circuit. The frequency of oscillation is determined solely by the crystal. The oscillator is a sine-wave generator and is ordinarily followed by a buffer amplifier, not only to supply additional gain, but to isolate the crystal oscillator circuit from the load. The purpose of the oscillator is to act as a stable frequency source in RF applications.

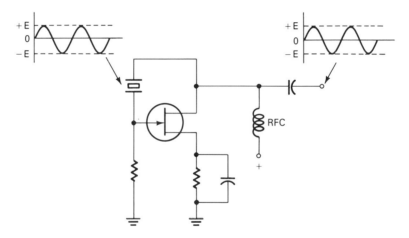

Figure 8.23.   Pierce crystal oscillator.

NOTCH FILTER

Depending on whether a receiver is AM, FM, or TV, the band of frequencies to be passed by IF stages will extend from relatively narrow to quite wide. The wider the band, the more likely it is that some persistent interfering signal of large amplitude and having a frequency somewhere within the IF range will be present.

A tunable notch filter (Figure 8.24) is sometimes used to eliminate the unwanted signal. As shown in the figure it consists of an L/C network tuned to the interfering frequency. At that frequency this tuned circuit has an extremely high impedance but with the impedance decreasing sharply on either side. Known as a "notch filter," once the tuning is correctly adjusted the circuit requires no further attention.

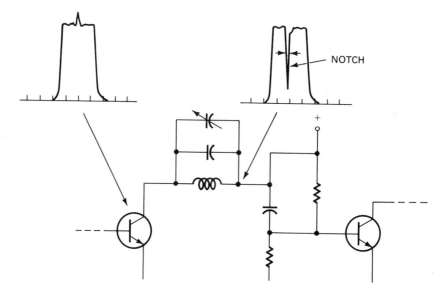

Figure 8.24. Notch filter.

SAWTOOTH GENERATOR

A circuit for generating sawtooth waveforms can consist of nothing more than a neon lamp, a capacitor, and a variable resistor, plus a source of DC voltage (Figure 8.25). The sawtooth frequency is in the audio range with an upper frequency limitation of about 5 KHz. The operating frequency is determined by the values of the variable resistor, R1 and capacitor C1, with various frequencies obtained by the setting of the potentiometer.

When the circuit is first turned on, the capacitor gradually charges, producing the trace ramp of the sawtooth. At some point during its charging time the voltage across the capacitor reaches the ionizing potential of the neon lamp. With ionization that lamp changes from a nonconductor to the equivalent of a low resistance, giving the capacitor a chance to discharge. This produces the retrace ramp of the waveform. The capacitor continues to discharge, and as it does so, the voltage across it drops sharply and finally reaches the point where the neon bulb can no longer ionize, becoming nonconductive. The capacitor then starts to charge again and the entire process repeats.

The setting of the potentiometer controls the frequency of the sawtooth waveform since it determines the speed with which the capacitor charges.

While the neon tube oscillator is simple, it has a number of faults that make it impractical for present-day use. The ionization point of the tube isn't precise, and the tube is capable of producing electrical noise. While potentiometer R1 can select the sawtooth frequency, frequent readjustment may be required to help maintain that frequency.

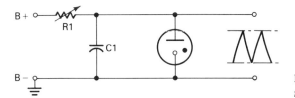

Figure 8.25. Neon tube sawtooth wave generator.

A better arrangement is the circuit shown in Figure 8.26. The basic idea of using the charge and discharge of a capacitor is retained and is provided by series resistors R7 and R8 shunted across capacitor C1. The time of charge and discharge of the capacitor is controlled by Q1 working in a blocking-oscillator circuit. The frequency of the output sawtooth wave is controlled by input sync pulses delivered via R1.

When the unit is first turned on, the transistor is in its cutoff condition. A small current does flow through the primary of the feedback transformer, T1. This induces a small voltage across the secondary winding, producing an equally small amount of forward-biasing voltage. This in turn increases the collector-emitter current, the start of an action that continues to grow. Current flowing through resistors R7 and R8 starts to charge capacitor C1, ultimately reaching maximum. The transistor is taken out of this condition by the application of an incoming sync pulse, and transistor conduction rapidly reaches cutoff with the shunt capacitor C1 discharging through R7 and R8.

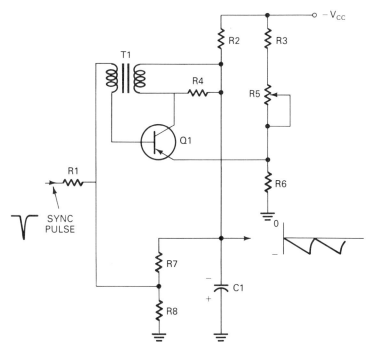

Figure 8.26. Blocking oscillator sawtooth generator.

ANTIQUE RADIO CIRCUITS

Electronic circuits were developed almost simultaneously with communications equipment since the need for a graphic explanation was immediately obvious. The first method was probably pictorial, and while this had advantages, was slow and time-consuming. Symbols were not only faster but could be drawn by anyone.

Figure 8.27A shows a diagram of an early radio receiver, and in this example consists of just two parts: a crystal detector and a headphone set. The headphone set shows two earpieces, but earlier receivers had just one. Circuit diagrams were usually accompanied by pictorials (Figure 8.27B) since the symbols alone were not enough to aid would-be constructors. This receiver had some advantages. It was easy to build and could be constructed quickly. It was an inexpensive and exciting development for a public intrigued by the idea of getting music "for nothing."

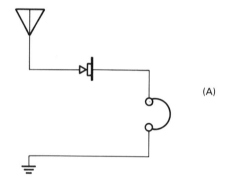

(A)

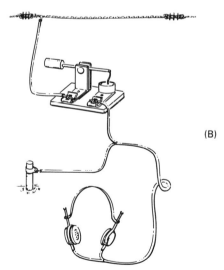

(B)

Figure 8.27. Circuit diagram of crystal radio receiver (A); and pictorial (B).

Further, the receiver had no on-off switch and required no battery or line power, consequently costing nothing to operate. It had a number of deficiencies, though. While it was suitable in areas that had just a single broadcasting station, its lack of tuning meant difficult listening with more than one station active. The output volume was low, so group listening was impractical.

The circuit was improved by adding a tuning coil (Figure 8.28A) that included a pair of sliders, permitting selection of various amounts of inductance. The capacitor shunted across the headphones was intended to separate the higher carrier frequency from the audio signal that passed through the headphones. As usual, a pictorial diagram (B) accompanied the circuit diagram.

With the arrival of the vacuum tube, crystal detectors went into a decline. An early receiver used a vacuum-tube diode or a triode connected as a diode (with the control grid connected externally to the plate of the tube) as a replacement for the

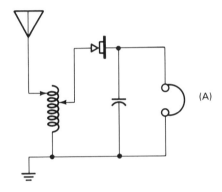

(A)

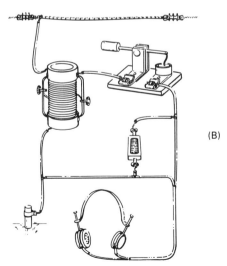

(B)

Figure 8.28. Crystal set with tuning coil (A); pictorial (B).

crystal detector (Figure 8.29). This had no advantage over the crystal and actually had a number of disadvantages. Initially, dry cells were used, but since they had a short operating life were soon replaced by a storage battery, sometimes called an "A" battery, to heat the tube's filament. The battery, quite similar to today's car battery, required recharging at various intervals and was large and messy for in-home use. Further, the tube diode did not supply amplification and so was no better in this respect than the crystal.

There was one important improvement, however, and this was the use of a variable capacitor across a coil connected between antenna and ground. It was easier

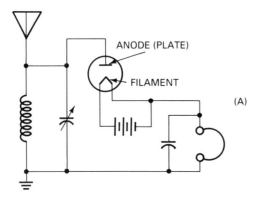

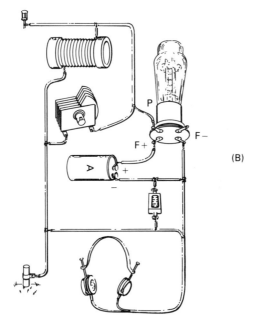

Figure 8.29. Early radio receiver using vacuum-tube diode (A); pictorial (B).

to tune than the slider type, but selectivity remained quite poor. Figure 8.29B is the pictorial for this receiver.

When the triode finally arrived, it gave radio (and electronics in general) a tremendous push. The triode (Figure 8.30A) supplied not only signal amplification

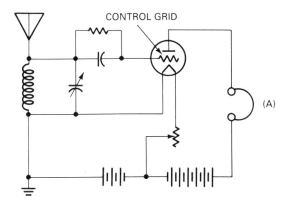

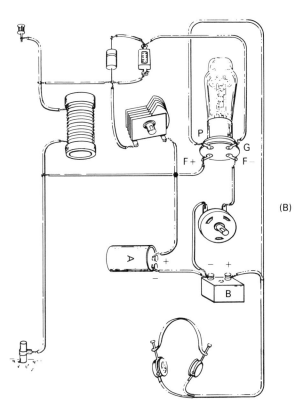

Figure 8.30. Early radio receiver using triode vacuum tube (A); pictorial (B).

but greater sensitivity for signal pickup. The receiver had its disadvantages, for initially it required three batteries: an "A" battery for heating the tube's filament, just as in the case of the diode, an anode (then more often referred to as a plate) or "B" battery, and a bias or "C" battery. Selectivity was no better than it had been with earlier radios, but with improved sensitivity antennas could be made smaller. Figure 8.30B illustrates the pictorial. Ultimately the "C" battery was eliminated, and a grid leak resistor and shunting capacitor were used in its place.

A number of events that affected circuits took place in fairly rapid succession. To increase volume more tubes were added to work as audio amplifiers, leading to the use of speakers. More tubes were also included, preceding the detector. These were tuned-radio-frequency amplifiers and increased selectivity substantially although it would be considered quite poor by today's standards. Each of the stages had to be tuned separately, so selecting a wanted station meant making three tuning adjustments, repeatedly, until interfering stations were tuned out or minimized. The receivers were known as tuned radio frequency or TRFs, subsequently replaced by the superheterodyne. Since its inception more than a half-century ago, this latter circuit has been the most widely used, not only for radio, but for television as well.

As receivers became more and more complex, so did their circuit diagrams. Some of the symbols used then gradually disappeared, while many of them were modified into the forms known today. Symbol changes, though, were quite modest, so a knowledge of present-day symbols should make reading antiquated diagrams easy.

9

How To Read
Electronic Graphs

In electronics we are often concerned with two or more variables. If we want to know the frequency response of an amplifier, we must compare the output signal amplitude against a frequency range extending from 20 Hz to 20,000 Hz. This can be done by preparing a list of what the strength of the signal would be at a list of frequency points. If we selected 100 Hz as the position for each of those points, we would have a list of 200 points, not only long and cumbersome, but mostly useful for the points that were selected. A better arrangement is to have a graph: it not only occupies less room but shows the frequency response at a glance.

LOCATING A POINT IN SPACE

If you were to pencil a dot at random on a page and wanted to identify its location, you could specify it by measuring its distance from the left edge of the paper and also its distance from the bottom edge. These two edges would then act as references.

A comparable method, but somewhat easier, is to use paper that has been pre-printed with a series of horizontal and vertical lines, dividing that paper into tiny squares. Such paper is available with 5 squares to the inch, 10 squares to the inch, and so on. The larger the number of squares to the inch the smaller they become, but in all instances the squares retain their geometric shape, that is, each of the four sides of each square have the same dimensions. In that case no measurements from the edges would be needed and the dot could be located by the square in which it was positioned. A sheet of paper of this kind is known by the general name of graph paper, but since there are a number of different kinds it is better to use its specific title. Such paper is variously known as Cartesian coordinate, quadrille, rectangular grid, or simply as "square" graph paper (Figure 9.1).

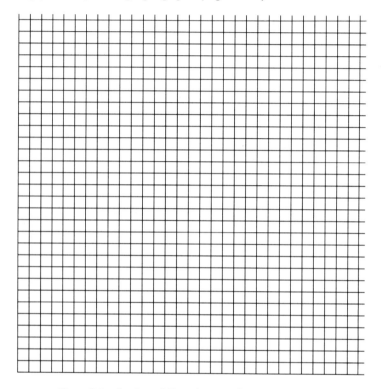

Figure 9.1. Section of Cartesian coordinate graph paper.

The graph paper can be divided into four large areas of equal size by drawing a vertical line called a Y-axis or ordinate dividing the sheet into two equal rectangles. A horizontal line referred to as an X-axis or abscissa can be drawn so that the page now consists of four squares of equal area. Each of these squares can now be named, moving in a counterclockwise manner, as quadrant 1, quadrant 2, and so on. The point at which the two axes cross is labeled zero (Figure 9.2).

GRAPH NUMBERING SYSTEM

Graph paper that has been prepared in this way is known as linear, since each of the tiny squares have the same dimensions. We can number alternate horizontal and vertical lines on the sheet but do so only in a linear manner, that is, the numbering

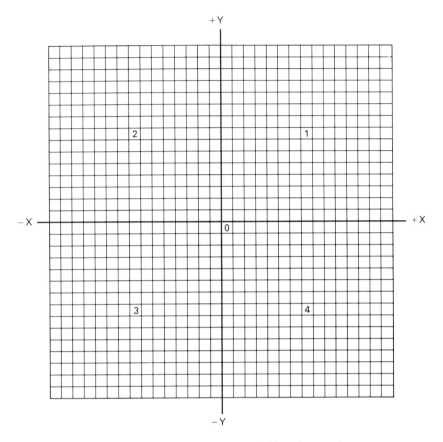

Figure 9.2. Square graph paper, divided into four quadrants.

system we select must also be linear. Thus, if we identify the first square along the X-axis to the right of the 0 mark as 1, the next square must be 2, and not some intermediate value as 1.37 or 11, or some similar number. The third square would be 3, the fourth would be 4, and so on. Each individual square does not need to be numbered. We could use a numbering system such as 0, 5, 10, 15, and so on.

Not only can we number along the horizontal axis but the vertical axis as well. In all instances, though, we start at the central or 0 point. Numbers to the right of that point are considered plus, although the plus sign is omitted. Numbers to the left of the 0 point along the X axis are preceded by a minus sign. A similar arrangement takes place along the Y axis. Divisions above the 0 point are considered plus; those below it as minus. Usually, though, the plus signs are omitted. As far as the numbers for each block placed to the left and right, above and below the 0 point, these are mostly omitted, with just enough remaining so we can identify the number value of each block. Further, the identification of the axes as X and Y is also eliminated, even though, in a description of the graph there might be some reference to the X axis or Y axis. The plus and minus signs have nothing to do with addition or subtraction and in this case simply indicate direction.

We can now use this system for locating any point on the page without a ruler. A point such as 4X means a point along the X axis, or four blocks to the right of zero. A point such as − 5x is also along the same axis but to the left of zero.

We can also move in a vertical direction. 2Y means a point that is two blocks up on the Y axis and a point such as − 3Y is a point on the Y axis three blocks down. All of these points, though, are along a single axis, either X or Y, and as such require a single number only preceded by either a plus or minus sign. Again, the plus sign is usually omitted.

To locate a point anywhere in the space occupied by the four quadrants, a pair of identifying numbers known as coordinates is needed. The first number of the pair is always the X-axis number; the second the Y-axis.

Using this technique a point can be located in any of the four quadrants. The first pair of coordinates, for example, could be 3, 4 (Figure 9.3). Move out to the

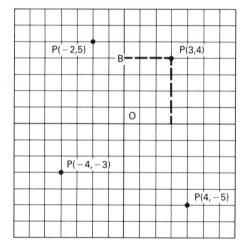

Figure 9.3. Method of locating points in each of the four quadrants.

right to 3 along the X axis and then up to 4, thus supplying point P in the first quadrant. The dashed lines shown in the illustration are simply guide lines from the X and Y axes and have no other significance.

The next pair of coordinates is − 2, 5 and the point is located by moving 2 units to the left along the X axis and 5 units upward on the Y axis. The result is a dot identified as P and is located in the second quadrant. The next pair of coordinates is − 4, − 3. As usual, the first of these two numbers is along the X-axis, the second along the Y-axis. The minus signs indicate a movement to the left along the X axis and then down on the Y axis. Point P shows the location of the dot indicated by the coordinates. The next pair of coordinates is 4, − 5 and is represented by P in the fourth quadrant.

For coordinates, then, a minus sign means to move to the left of the center

0 point along the X-axis, or down on the Y-axis, while a plus sign (or the absence of a plus sign) means a move to the right along the X-axis or up on the Y-axis.

FROM POINTS TO LINES

A line is a succession of points positioned so closely to each other that it is impossible to determine the space between them. However, it isn't necessary to identify each dot in that line to be able to draw that line.

As an example, consider the graph in Figure 9.4. We have been supplied with three pairs of coordinates: 2,4; 3,5; 4,6. Since all of these coordinates are positive, we know the line we want to draw will be in the first quadrant, since this is the only quadrant in which all the coordinates are positive. Each of the coordinates results in a point on the graph, and since there are three pairs of coordinates we will then have three points. The coordinates for each of these points can be indicated on the graph simply to emphasize point location, but normally this information is not included.

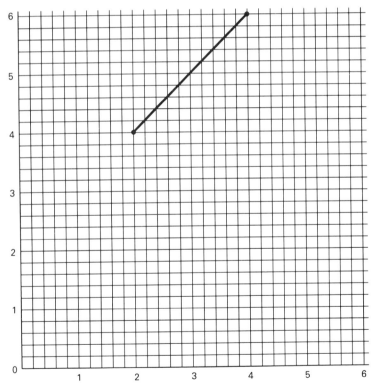

Figure 9.4. Graph is a line produced by connecting points representing coordinates.

We can connect the three points and the result will be a straight line and that line is a graph of the coordinates that were supplied.

Quite often a graph will have just a single straight line, but if we are supplied with enough coordinates we will have a number of such lines. Further, although the individual lines in Figure 9.5 are straight, using just two pairs of coordinates, it is also possible for a curve to appear when a larger number of coordinate points are connected.

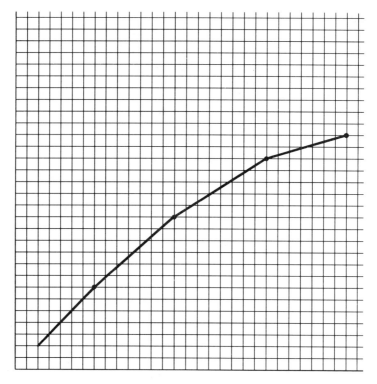

Figure 9.5. A single pair of coordinates can result in a straight line; a succession of coordinates can produce a curved graph.

PLOTTING A GRAPH

The numbers marked along the horizontal and vertical axes of a graph are known as scales. The X axis might be calibrated in terms of time; the Y axis in units of temperature (Figure 9.6). Note that this does not violate the requirement of linearity. In this example, each hour on the X axis occupies the same amount of space. Each degree of temperature along the Y axis has the same amount of space as any other degree.

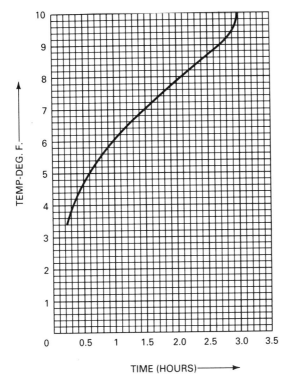

Figure 9.6. Graph of temperature vs time.

The action of producing a graph from a collection of coordinates is called "plotting." Technically, to plot a straight line graph just two pairs of coordinates are required, but greater accuracy can be obtained by using more than just these. Note that the graph in Figure 9.5 is a curve, but this curve consists of a series of straight lines, each of which has been produced by plotting pairs of coordinates.

The advantage of the graph in Figure 9.6 is that you can immediately see the effect on temperature with the passage of a time period of 3½ hours. You might be able to deduce this by examining a report of temperature over this time span, but a graph tells the whole story visually.

ELECTRONIC GRAPHS

The gain of an amplifier can vary with frequency, as indicated in the graph in Figure 9.7. Here the horizontal axis is marked off in MHz, while the voltage gain is indicated along the vertical axis. Since all the numbers are positive (plus), plotting the coordinates takes place in the first quadrant only. The graph shows that the gain peaks a little above 3 MHz, and from that point gradually decreases to zero.

Electronics graphs are often used to supply a graphic display of the relationship between two quantities. The curve in Figure 9.8 indicates the amount of base

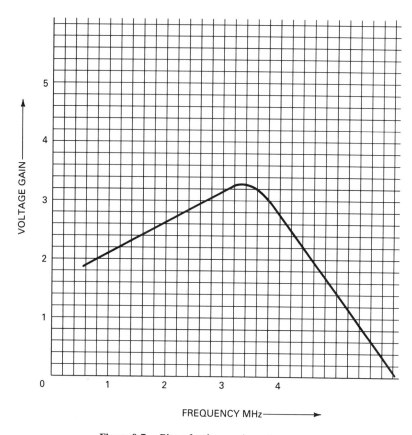

Figure 9.7. Plot of voltage gain vs frequency.

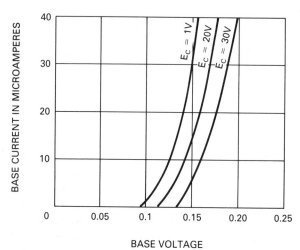

Figure 9.8. Family of curves.

current in microamperes for a transistor used in a common-emitter circuit. The collector voltage is held constant at 30 volts, with this data marked alongside the curve. The graph shows that as the base voltage is increased by very small amounts, the base current increases moderately at first and then more sharply.

It is also possible to use the same graph to produce a series of curves under different operating conditions. Thus there are two additional curves, one for a constant collector voltage of 20 volts and another for a collector EMF (electromotive force) of only 1 volt. A group of curves drawn in this manner is known as a family of curves. Such families are quite common in electronics since they are useful in making comparisons between different sets of operating conditions.

THE CARTESIAN COORDINATE PROBLEM

When plotting a single graph or a family of graphs, it is helpful if the values of both sets of coordinates are close to each other arithmetically. Plotting a graph for 2,3, 4,5, and 7,10 isn't difficult since there isn't much of a difference between the numbers. But if we have coordinates such as 2400, 3985 or 42 2000, producing the graph would be quite difficult since the Y numbers would require a rather long sheet of paper. A graph of this kind would be impractical.

As a further example consider a sheet of Cartesian coordinate graph paper having 20 squares to the inch. Also assume that each horizontal line has a length of 7 inches. If this sheet was being used to plot voltage against current, each inch along the bottom could represent 20 volts. The maximum voltage that could be indicated would be $7 \times 20 = 140$ volts. But if some of the voltage values were small, such as 3 or 5 volts, with others much higher, such as 210, 350, and 890 volts, this graph paper would not be suitable.

But suppose that the voltages plotted along the horizontal axis were all less than 140 volts. It is possible that the current values being plotted along the Y axis would be so large that they would take the graph right off the sheet of paper.

THE LOGARITHMIC GRAPH

Figure 9.9 illustrates two horizontal lines. The upper line is marked off linearly, that is, the distance from one point on the line to the next is always the same, and in this case is 1 MHz.

The second horizontal line is known as logarithmic. The first division at the left is marked 1000 Hz while the next is 10,000 Hz, the third 100,000 Hz, and so on. Now consider these numbers in terms of logarithms (logs). The logarithm of $10 = 1$; log $100 = 2$; log $1,000 = 3$, and so on. Note that the two lines have almost the same length but that most of the bottom line involves frequencies from 1000 Hz to 1 MHz. The upper line, though, has a much shorter distance to 1 MHz.

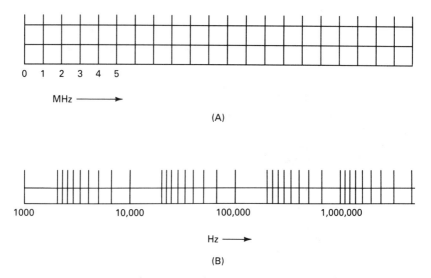

Figure 9.9. Linear scale (A); logarithmic (B).

The Semilogarithmic Graph

It is possible to have the coordinates to be used in plotting a graph to have extreme variations in both the X and Y directions. However, quite often the coordinates along the X axis are numbers that are reasonably close to each other, while the Y coordinates have an extremely wide range. In that case we can use semilogarithmic graph paper for plotting the curve (Figure 9.10).

The X axis can be any horizontal line extending completely from left to right. This axis is linear and is divided into equally spaced blocks. This means that one set of coordinates, the X coordinates, can be accommodated. However, if we move vertically you can see that the vertical divisions are arranged in logarithmic form. A preprinted sheet of graph paper of this kind is known as semilogarithmic since it is linear in one direction (either horizontally or vertically) and logarithmic in the other. It is called semilogarithmic since only one axis is logarithmic.

In some case the numbers along the Y axis cannot be accommodated and so the graph can be extended (Figure 9.11). Since there are now two logarithmic sections, the graph paper is referred to as two-cycle semilogarithmic. We can carry this concept one step further and obtain three-cycle paper.

Now consider the numbers that are printed alongside the left side of the sheet (Figure 9.12). The numbers are shown as 1 through 10 for the first cycle, 1 through 10 for the second cycle, and again, 1 through 10 for the third cycle. Assuming we start with 10, then the next number shown as 2 is actually 20, the following number, 3, is 30, and so on. This means that 1 at the end of this group of numbers is 100. But 100 is the beginning of the second cycle. We then have, in succeeding order, 100,

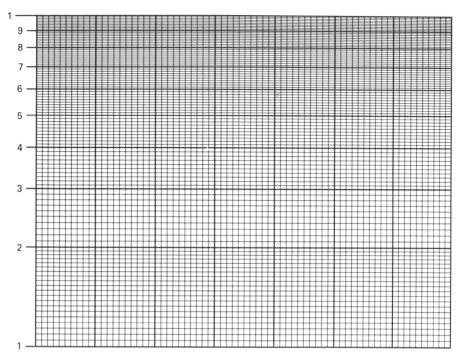

Figure 9.10. Semilogarithmic graph paper. The X-axis is linear; the Y-axis is logarithmic.

200, 300, and so on. In this second cycle the number 8 corresponds to 800. Consequently, the number 1 at the end of the second cycle is equivalent to 1,000. But if we start the third cycle with 1,000, the number 2 in that cycle is the same as 2,000, and the number 10 at the top left-hand side is equivalent to 10,000. In other words, we now have a graph which in the vertical direction extends from 10 at the bottom to 10,000 at the top. Across the bottom we can write 1, 2, 3, and so on, until we reach digit 7. Thus, with this graph we can plot any X coordinate having a value from 1 through 7, and any Y coordinate having a value from 10 through 10,000. This should be adequate for most problems in graphing.

Of course, it is entirely possible that even this wide numbering range isn't enough, and so in that case we can use four-cycle semilogarithmic graph paper. Since with three-cycle paper we ended with 10,000, the next value along the Y axis is 20,000, ending with 100,000 at the top.

We have no assurance, though, that all the numbers along the X axis will be small numbers, closely related. It is possible that we will have a wide number variation along both axes. In that case we can use graph paper that is logarithmic in both

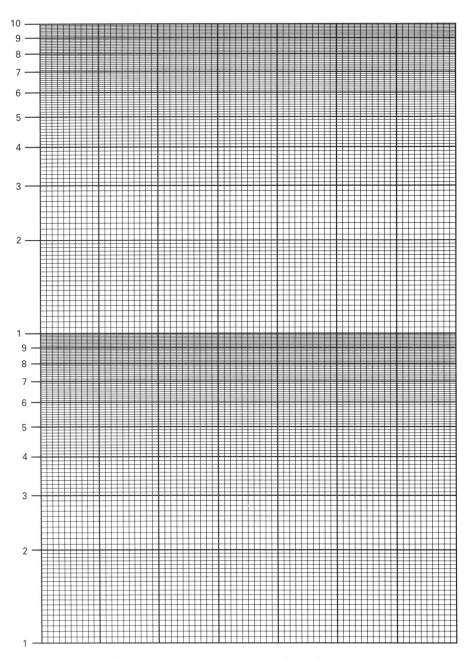

Figure 9.11. Two-cycle semilogarithmic graph paper.

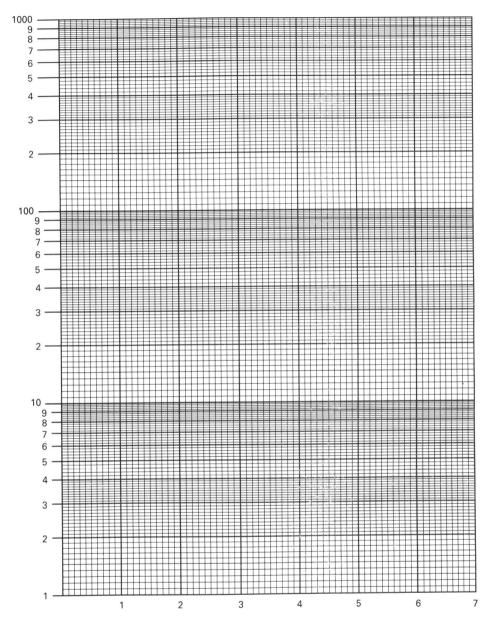

Figure 9.12. Three-cycle semilogarithmic graph paper.

directions (Figure 9.13). If you will examine the drawing you will see it is three cycles vertically and also three cycles horizontally. This type of paper is known as three-by-three-cycle logarithmic graph paper or log-log paper. Of course we can have variations, such as one-cycle by one-cycle, and so on.

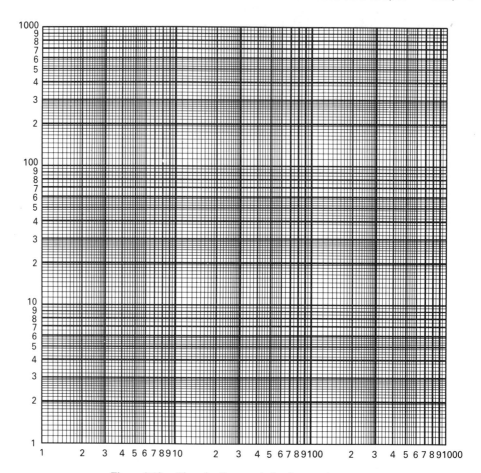

Figure 9.13. Three-by-three cycle log-log graph paper.

Numbering the Scale

When working with a number of X and Y coordinates it is essential to examine them carefully to determine just which type of graph paper would be most suitable. The next step would be to decide on a suitable scale. If, for example, all of the values for the Y coordinate were decimal values, then the largest number would be 1. A series of Y coordinates, such as .04, .2, .5, would supply a good clue as to the scale to be used. Such a scale is shown in Figure 9.14A, representing a partial sheet of two-cycle semi-logarithmic graph paper.

If, however, the Y coordinates were all somewhere in the region between 2 and 200, we could set up a Y scale as indicated in Figure 9.14B. However, if the Y coordinates were between 10 and 1,000 the scale could be as in C. Note that the same graph paper was used for all three conditions with the only difference being the scale that was selected.

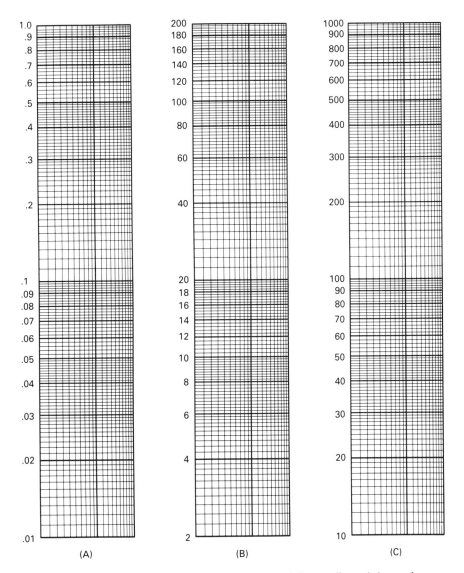

Figure 9.14. Scale selection depends on the values of the coordinates being used.

A variety of different kinds of graph paper is available. Thus, the logarithmic scale can be vertical or horizontal. In Figure 9.15, the horizontal scale is linear; the vertical scale is logarithmic. Further, the graph shows four logarithmic cycles. With this sheet of graph paper we can plot numbers along the vertical axis from one to 10,000. Along the horizontal axis we can plot numbers from 1 to 8.

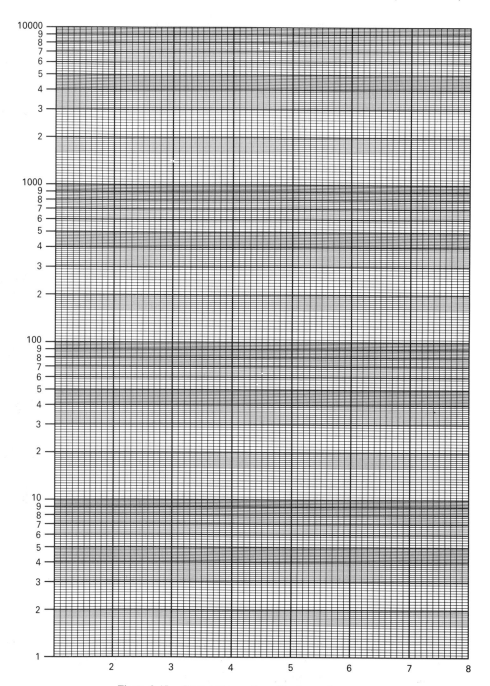

Figure 9.15.   Four-cycle semilogarithmic graph paper.

REFERENCE LINE

Ordinarily, the baseline is the reference or zero line, but any line can be selected for this purpose. If the resulting graph is one that shows both positive and negative values, the position of the reference line can be adjusted, with values above this line indicated as positive, those below it as negative (Figure 9.16). In this graph, the vertical divisions are linear and the horizontal divisions are logarithmic. The vertical divisions have a range of plus 15 to minus 15, while the logarithmic scale extends from 20 Hz to 20 KHz. A logarithmic scale was selected in this example because of the wide disparity between these two number groups. The graph paper used is three-cycle semilogarithmic.

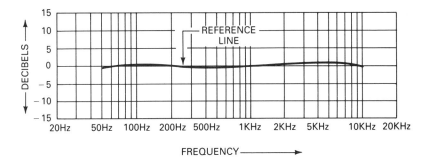

Figure 9.16. Positive and negative values plotted on semilogarithmic graph paper.

THE SCHEMATIC GRAPH

To convey the greatest amount of information, a schematic diagram is sometimes made part of a graph (Figure 9.17). Since the schematic intrudes on the graph area, it must be small. If inclusion of the graph isn't possible, it can be placed adjacent to the diagram.

THE BALLOON GRAPH

As in the case of using balloons in block diagrams to supply more information than can be contained in a block, so too is it possible to use balloons to detail data about parts of a curve. Balloons can be used to call attention to the significance of certain parts of the curve. As an example, the frequency response of an audio amplifier can be flat from 20 Hz to 20 KHz with a variation of one-half dB, plus or minus. Deviations indicate some fault in the amplifier. The results of such variations can be emphasized through the use of balloons (Figure 9.18).

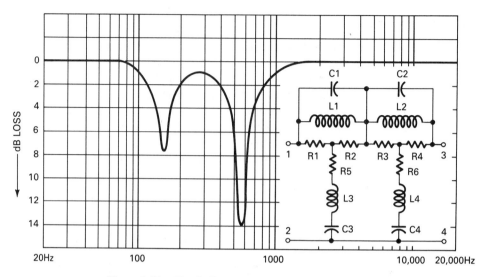

Figure 9.17. Circuit diagram is sometimes included with graph.

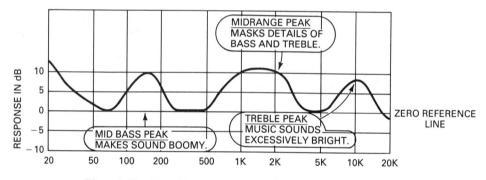

Figure 9.18. Use of balloons to emphasize certain parts of a graph.

THE MULTIGRAPH

As indicated previously in Figure 9.8, a graph can show a family of characteristics of a component working under different operating characteristics. A graph can also be used to plot a number of different characteristics (Figure 9.19), and in this example we get a contrast between power gain, voltage gain, and current gain. Since we are plotting one characteristic, load resistance vs. three different characteristics (power, voltage, and current), we require three different Y axes or ordinates, one for each of the characteristics. The alternative would be to produce three graphs instead of the one shown. The disadvantage would be that the voltage, current, and power relationships wouldn't be shown as effectively.

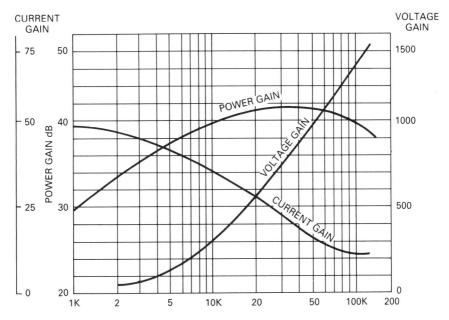

Figure 9.19. Multigraph.

THE BAR GRAPH

A graph is plotted by determining the position of a number of points on a sheet of graph paper and then connecting them. However, instead of connecting the points, a vertical line can be drawn between each of them and the base line (Figure 9.20). If these lines are sufficiently close to each other, the curve their end points produce can be visualized. A graph of this kind is suitable when only discrete values are of interest, and not any intermediate values.

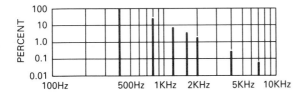

Figure 9.20. Bar graph.

THE ZERO POINT

The various graphs that have been illustrated are nearly all made using the first quadrant. The number 0 was selected as the crossing point for both axes, X and Y. However, this assumes 0 is the starting point for both the ordinate and the abscissa.

For some graphs this might be suitable, but not for others (Figure 9.21). This drawing is a log-log graph. For the Y axis the starting point is 0.1 and is .01 for the X axis. If the wrong starting points are selected, it is possible for a curve to appear at the top or bottom of the graph or some other awkward location. It is desirable, as shown in this illustration, for the graph to begin directly at the Y axis.

However, the final positioning of the graph is optional. By starting at the extreme left you have more of an assurance that the graph will not run off the page at the right. From an esthetic point of view it might be desirable to have the graph centered on the page.

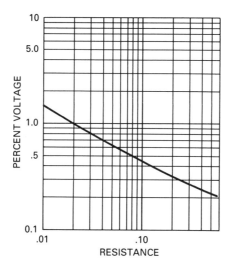

Figure 9.21. Log-log graph using two different starting points.

COORDINATE POINTS

The appearance of a graph will also depend on the selected values of coordinate points and whether the graph is plotted on Cartesian coordinate or on logarithmic paper. If, for example, one of the variables being plotted is current in terms of milliamperes, selecting values of amperes will result in a graph so small as to be unreadable. Thus, we have two possible graph conditions: one in which the graph is too small to be of use, the other in which the graph is too large for the page. There is also the need to determine whether to use linear paper, semi-logarithmic, log-log, or single-cycle, double-cycle, etc.

COORDINATE SELECTION

Plotting a graph is a comparison between a pair of variable values. Either one of these values can be used for the X axis, with the other intended for the Y axis. If each set has the same number of values it will usually make little difference which one is chosen to represent the X axis or Y axis. But if one set has a much larger number, it is generally used to produce X axis points (Figure 9.22). In this drawing the X axis is so extensive that it requires a three-cycle logarithmic approach. The coordinates for the Y axis are few, and so the plotting for this axis is linear. Even if the X axis can use the linear plotting form, the larger number of coordinate points determines the selection of a three-cycle log.

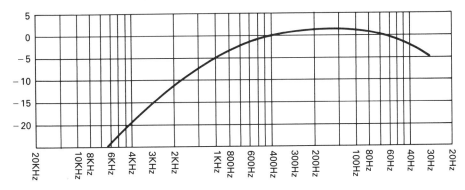

Figure 9.22. Graph having few coordinates for the Y-axis; extensive numbering for the X-axis.

THE FOUR-QUADRANT GRAPH

Nearly all graphs are single-quadrant types with all coordinate points for the X axis and Y axis positive. In some instances though, a four-quadrant graph (Figure 9.23) is essential. In this example we have two sets of X and Y values. In the first set all

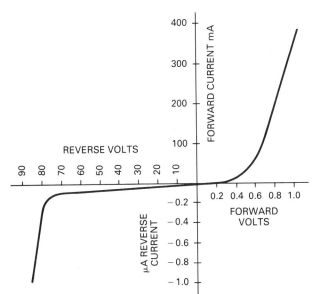

Figure 9.23. Linear graph using first and third quadrants.

the coordinate points are positive; in the second set they are all negative. In this example only the first and third quadrants are used, and while the second and fourth are blank, they are included to emphasize the relationship of the positive and negative points. This graph shows the characteristic of a silicon junction diode.

PLOTTING DECIMAL VALUES

It is possible to plot X and Y values when each of these is in decimal form. Similarly, we can plot them when one is in decimal, the other in whole numbers (Figure 9.24). In this example, the Y axis is decimal; the X in whole number form. The problem with using decimal values is that these sometimes require approximations. If the Y coordinates in this drawing had values of 0.1; 0.2; 0.3, and so on, it would be easy enough to determine the position of these points along the Y axis. But when these are 0.3011; 0.478; 0.602, and so on, placing the points is simply an educated guess. Further, when decimals are supplied having three or four decimal places, there is an implication of great accuracy, an implication that is misleading.

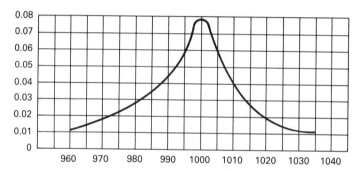

Figure 9.24. Linear graph plotting whole numbers against decimal values.

THE GENERALIZED GRAPH

It is possible to have a graph in which no coordinates are available. The purpose of such a graph is not to give specific information about the behavior of a circuit using definite values of X and Y, but rather to supply a curve that will just be an approximation (Figure 9.25A). A curve such as this could serve in a classroom or possibly in anticipation of a curve that is more detailed. It could also be used as a rough graph so as to decide on the number of quadrants required for a more formal approach.

The difficulty with the generalized graph is that there is no way of knowing whether it is linear or logarithmic. All that it does is indicate the functioning of a pair of variable quantities.

Figure 9.25B indicates another type of generalized graph but with more details supplied. Values are indicated along the X and Y axes, and because of their spacing

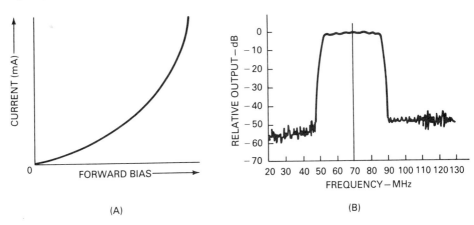

Figure 9.25. Generalized graph (A). More detailed generalized graph (B). This graph shows the frequency response of a SAW filter in an IF stage.

it is obvious that this graph is linear. However, the vertical and horizontal lines of a Cartesian coordinate graph have been omitted. This is a common practice since these lines, essential when the graph is being plotted, aren't always required for reading the values of the graph.

GRAPHING A SINE WAVE

Any circle, regardless of its diameter, can have its circumference divided into 360 equal parts or degrees (Figure 9.26A). If this circle is made of a spring-like material and is cut at one point, we will then have a line divided into 360° (Figure 9.26B). We can use this line as the basis for graphing a sine wave.

This line has been divided into 45° segments, starting with 0° and ending with 360°. From each of these we erect a vertical line. Adjacent to this line we have a circle with a rotating vector, a line representing the rotation of an armature in an AC generator (Figure 9.27). When the line is horizontal, the armature does not cut any of the magnetic lines of flux of the magnet surrounding the armature. This is the 0° degree position. But as the armature rotates it passes through more and more flux lines, until at 90° it intercepts the maximum number. As the armature continues rotating, approaching the 180° point, it cuts fewer and fewer lines. Following 180° the process repeats.

From the rotating vector we can project horizontal lines. Thus a line from the 45° point will intersect the 45° vertical line from the X axis (Figure 9.28). If we repeat this procedure at all the selected degree points shown on the circle, we will have a series of points. Connecting these results is a sine wave, one of the most common waveforms used in electronics. The lower half of the sine wave is the mirror image of the upper half. It simply represents the negative half of the sine wave, or a reversal of current in the armature coil.

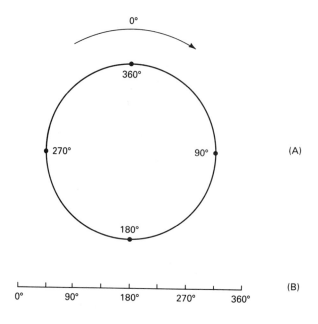

(A)

(B)

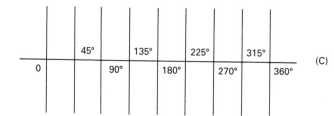

(C)

Figure 9.26. Every circle (A) or straight line (B) can be divided into 360°. Series of Y ordinates of random length erected along 360° line (C).

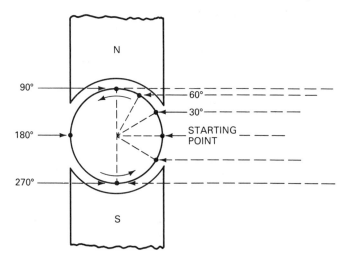

Figure 9.27. Coil rotating in the magnetic field of a magnet, develops a voltage. Horizontal projection lines, each an abscissa, are projected from various positions of the coil.

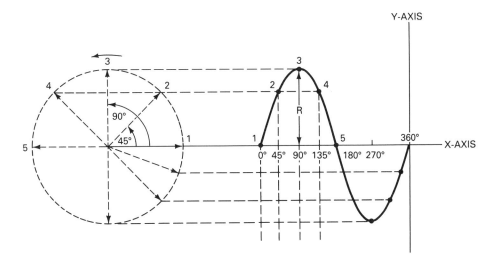

Figure 9.28. Intersection points of vertical and horizontal projections produces the graph of a sine wave.

The horizontal lines are equivalent to X coordinate values; the vertical lines Y coordinate values. The points along the X axis are linear; the Y values are non-linear. A more precisely drawn sine wave can be obtained by selecting 15° or 30° points along the X axis instead of 45°.

GRAPHS AND ELECTRONICS MATHEMATICS

Formulas are often used in electronics, and these are generally expressed as equations. One form of Ohm's law, for example, $R = E/I$, can be written this way or can appear as a graph. All that is necessary is to select values of current and voltage and use these as X and Y coordinates. The graph can then be used to learn the value of E for any corresponding values of I or the value of I for any corresponding value of E (Figure 9.29).

The graph is first plotted using figures for voltage and current as the coordinates. The result is the straight-line graph shown in Figure 9.29. This line represents a value of 5 ohms. It can then be used to find any value of current for any voltage ranging between 0 and 30. Thus, as indicated by the vertical and horizontal dashed lines, 20 volts would produce a current of 4 amperes. However, this graph is suitable only when the resistance value is 5 ohms. To emphasize this, the value of $R = 5$ ohms is marked on the graph. This assumes that the amount of resistance will remain constant, but this is not always the case, since resistors used in electronic gear have a positive temperature coefficient. Thus a graph of component or circuit behavior represents an idealized condition, subject to change based on operating conditions. This doesn't mean a graph isn't an electronics aid. It is, but it must be considered from a practical point of view.

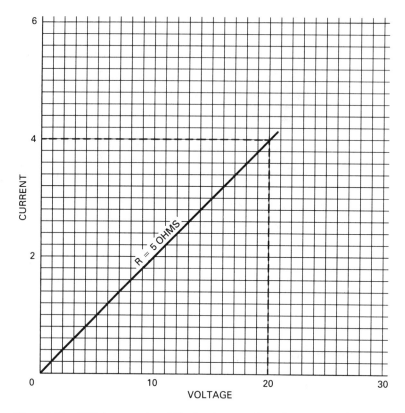

Figure 9.29.   With known value of fixed resistance values of current or voltage can be found without using a formula.

As an example, the resistance of the filament of an electric light bulb increases with temperature. Since the resistance is variable, the graph is a curve instead of a straight line.

SCALES

Any line, whether straight or curved, can be divided and subdivided into equal (linear), or into nonlinear or logarithmic units. A ruler is one example of a scale divided into English or metric units, inches in one case, centimeters in the other. A clock is a scale, completely circular, and is subdivided into hours and minutes. Either a ruler or a clock supplies a linear scale with the space between each division equal to any other space.

A nonlinear scale is one that has unequal divisions. The scale of an ohmmeter (Figure 9.30) is nonlinear and is more crowded at the left side of the scale, but gradually becomes more open in moving toward the right. Some scales have a reverse characteristic; they are less crowded at the left but have closer spaced scale divisions in moving toward the right.

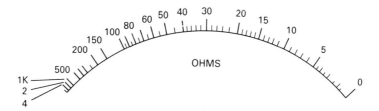

Figure 9.30. Ohmmeter scale is nonlinear.

Scales are a type of graph often used on meters. The three meter scales in Figure 9.31 are all linear, but they differ in the number of available divisions. Scales usually start with 0 at the left side, but there is considerable variation in the final number at the right. The one in Figure 9.31A ends with 1 at the right side with divisions in tenths. As long as the meter pointer rests on a division, whether that division is marked

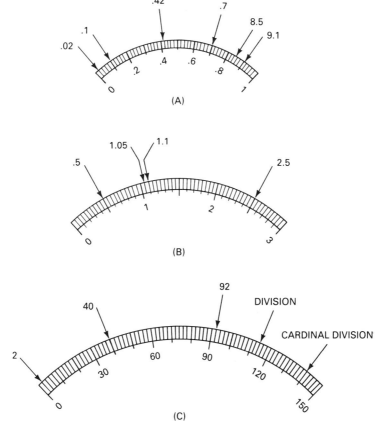

Figure 9.31. Linear scales. Drawing C has two units per division; 10 units per cardinal division.

or not, the specific value indicated by the meter is known, but points in between the divisions must be estimated, a process known as scale interpolation, often nothing more than an educated guess. This drawing shows meter readings of .02, .1, .42, and .7. Each one of these is produced by the meter pointer stopping at a division on the scale. The drawing also shows estimates of 8.5 and 9.1, but these are not as accurate as the other readings.

The scale in Figure 9.31B has more subdivisions, with each of these having a value of .05. The drawing shows points on the scale at .5, 1.05, 1.1, and 2.5. These are not interpolation points since no estimates need be made. However, points such as 1.13 or 2.53 would be between divisions and so would need to be estimated.

Figure 9.31C starts at 0 and ends at 150 on the right. Each division between those that are numbered has a value of 2. In this drawing the arrow points to 2, 40, and 92.

Errors are always possible when reading graphs or meter scales. One of these may be caused by using an insufficient number of coordinate values in the case of a graph that is a curved line, or, in the case of a meter scale by assigning incorrect values to unmarked divisions, or through parallax, that is, reading the scale from an off-center position, either to the right or left.

NOMOGRAPHS

A nomograph involves variables that depend on each other. A typical example of a nomograph appears in Figure 9.32. In this illustration we have a pair of adjacent scales, with one marked in degrees Fahrenheit; the other in Celsius. The conversion can be done by selecting the appropriate formula: $C = (F - 32) \times 5/9$ when the temperature in Celsius is unknown, or $F = (C \times 9/5) + 32$ when the temperature in Fahrenheit is unknown. To move from one temperature scale to the other, just put a straightedge horizontally across the graph.

The nomograph just described uses a pair of immediately adjacent scales with each of these involving two variables. The solution could be called single-step, for it involves nothing more than moving from one point on a scale, horizontally, to the corresponding point on the other scale. Sometimes, though, two scales will not do; three are needed.

The Four-Scale Nomograph

The formula for the power gain or loss of an amplifier or a network can be calculated from the formula dB = 10 log (P2/P1), and in this case, log means log to the base 10. An alternative method of solution is to use a four-scale nomograph (Figure 9.33).

While the formula does not involve much arithmetic, it does require determining the logarithm of the ratio of two powers. A simpler way, but one that isn't more accurate since it can require some interpolation, involves the use of a four-scale nomograph. Put a ruler across the nomograph so it underlines the amount of input power on the left scale and the output power on the right scale. If, for example, the ruler

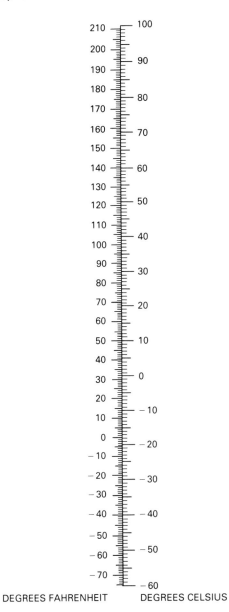

210	100
200	90
190	
180	80
170	
160	70
150	
140	60
130	
120	50
110	
100	40
90	30
80	
70	20
60	
50	10
40	
30	0
20	
10	− 10
0	
− 10	− 20
− 20	− 30
− 30	
− 40	− 40
− 50	
− 60	− 50
− 70	− 60

DEGREES FAHRENHEIT   DEGREES CELSIUS

Figure 9.32. Nomograph for temperature conversion.

is put under 0.1 on the left scale and 1.0 on the right scale, it will lie directly across the center scale from which it is possible to read both the voltage and decibel ratios. In this example it would be a power ratio of 10 equivalent to 10 dB.

The ruler need not be kept horizontal, but can be at any angle that permits reading the scales. A reading above the 0 mark on the dB scale means a power gain, below it a power loss, as indicated by the minus sign in front of the dB numbers.

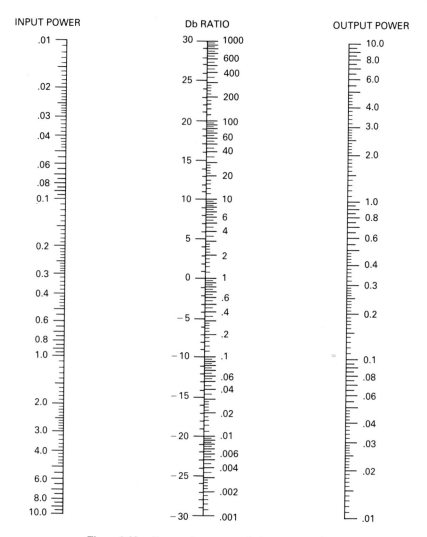

Figure 9.33. Four-scale nomograph for power ratios.

If the input power is 1.0 and the output power is .02, the power ratio is .02 and the loss in decibels is 17 dB. The nomograph can also be used to determine the amount of input and output power required for a given amount of gain or loss in dB. Put the ruler on the selected dB point and use it as a pivot point and read the amounts of input and output power from the respective scales.

Advantages of Nomographs

There are two advantages supplied by nomographs that the formulas on which they are based cannot supply. A nomograph can yield a large number of solutions, requiring nothing more than moving a ruler from one position to another. Also, the nomograph permits working backward from a given solution to the values that produced that solution.

The Tilted Scale Nomograph

Not all nomographs have scales that are parallel to each other. Figure 9.34 is a nomograph in which two of the scales tilt away from the center scale. This nomograph

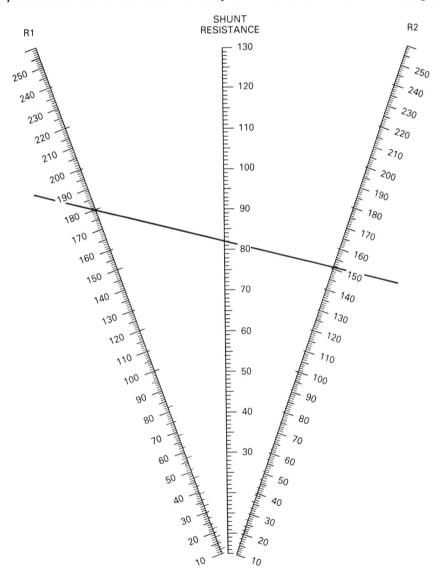

Figure 9.34. Nomograph for finding the equivalent resistance of two resistors, R1 and R2, in shunt.

is based on the formula $R = (R1 \times R2)/(R1 + R2)$, and is used for finding the equivalent resistance of two resistors in parallel. The two outer scales represent the known values of the two resistors. The center scale supplies the solution to the problem. As an example, the straight line in the illustration shows how to find the equivalent value of a 180-ohm resistor in parallel with a 150-ohm resistor. The straight-edge crosses the center scale at 82 ohms. The nomograph could also be used to find the equivalent of larger values provided they are in identical powers of 10.

Capacitors in series use the same formula as resistors in parallel, and so the same nomograph can be used. All three scales, though, must be in the same capacitance units: microfarads or picofarads.

An advantage of this nomograph is that you can use it to determine what resistor or capacitor values you need to obtain a desired amount of resistance or capacitance. If you have a supply of miscellaneous resistors but do not have a 60-ohm unit, put the straight edge on 60 on the center scale and use this as a pivot point. The straightedge, in cutting across the left- and right-hand scales, will supply a large number of possible resistor combinations. This technique can also be used for capacitors to be connected in series.

FOOTPRINTS

A footprint is a type of graph used in connection with satellite TV. A footprint (Figure 9.35) indicates the amount of signal strength received by different areas of continental U.S. from a specific satellite. The signal strength is measured and all points having the same amount are connected, resulting in an enclosed geometric figure, roughly resembling a circle. The numbers indicate signal strength in decibels referenced to one watt and written as dBw. In Figure 9.35, the center contour is 35 dBw.

The diagram is known as a footprint with the center of the innermost pattern referred to as the boresight point, and is the location of the strongest signal strength. In this illustration the outermost pattern has a signal strength of 32 dBw.

The footprint graph (or pattern) is not the same for all satellites, and often the pattern is different for the various transponders (transmitter/receiver) on the same satellite.

POLAR DIAGRAMS

A polar diagram is a specialized type of graph and consists of a number of concentric circles (Figure 9.36). Each circle is marked with a number representing decibels. The outermost is 0 dB, the next -5 dB, and so on. The numbers are negative so as to indicate a weakening of a sound. The graph is divided into quadrants, with one line running vertically from 0° to 180°, and the other horizontally from 90° at the left to 90° at the right. Each quadrant is bisected by lines extending from 45° to 135°.

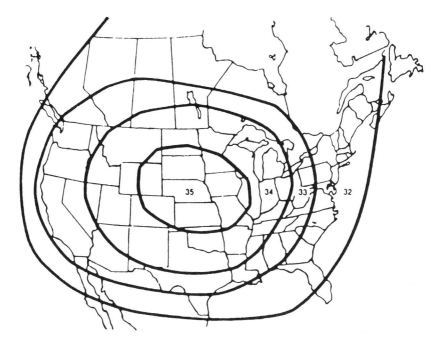

Figure 9.35.   Footprint (graph) indicating signal strength of a selected satellite.

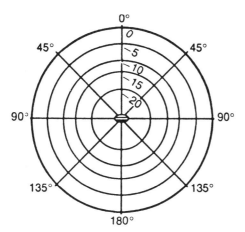

Figure 9.36.   Polar diagram.

While this drawing shows the basic elements of a polar diagram, polar coordinate graph paper (Figure 9.37A) is used since it is more detailed and permits drawing a more accurate graph. Such graphs are used for plotting the directional response curve of a microphone or the field pattern of an antenna (B).

For a microphone, a pattern is produced by a sound generator, usually working at a fixed frequency of 1 KHz while the microphone is rotated around it at a

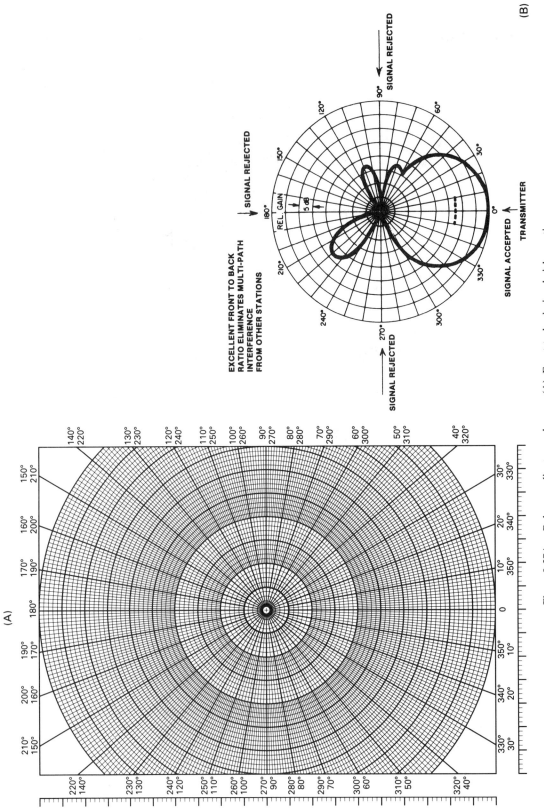

Figure 9.37A. Polar coordinate graph paper (A); Front-to-back signal pickup ratio of an antenna (B).

distance of 1 meter. This plane of rotation can be vertical with respect to the microphone, horizontal to it, or any other angular displacement. Two of the most common polar graphs resulting from this test are the omni (omnidirectional) and the cardioid. The omni pattern is circular while the cardioid is somewhat heart-shaped (Figure 9.38). This latter pattern shows that the microphone has optimum response on axis, that is, with the microphone facing the sound source directly. Sounds arriving from the sides are somewhat attenuated, while the least sound pickup is from the rear.

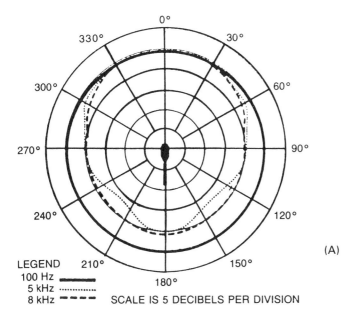

LEGEND
100 Hz
5 kHz
8 kHz
SCALE IS 5 DECIBELS PER DIVISION

(A)

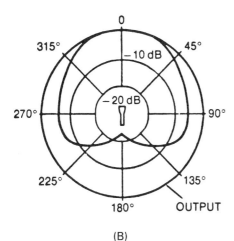

(B)

Figure 9.38. Microphone response patterns. Omni (A); cardioid (B).

GRAPH PRODUCTION METHODS

A graph can be drawn manually, using a selected type of graph paper. This is the least expensive method, but the most time-consuming. An oscilloscope can also be used, with the resulting patterns on the screen known as Lissajous figures. While the graph will disappear when the instrument is turned off, a record of the pattern can be made by using a camera. Special fixtures are made for holding the camera in place with its lens facing the screen of the oscilloscope. A computer can also be used and will supply the graph with the help of a printer.

Index